Advance Praise for *Fluke*

"A tour de force of masterful writing that weaves together simple and not-so-simple mathematical notions of probability and statistics into various intriguing coincidences from fact and fiction, explaining with nuance various strange phenomena. Mazur's book will teach you some of this mathematics, leaving you quite equipped to understand the role of chance in your life without resorting to magical thinking."
—Gizem Karaali, editor of the
Journal of Humanistic Mathematics

"The chances were very slim that you'd ever read this blurb. A simple-minded calculation puts the odds at about 50,000 to one against. Yet . . . here you are. How weird is this seemingly far-fetched coincidence? Well, dear reader, you've picked up the right book to answer that question."
—Charles Seife, author of *Zero:*
The Biography of a Dangerous Idea

"Joseph Mazur's *Fluke* walks the reader, hand in steady hand, through the weird and dangerous landscape of extreme probability, distinguishing cause from correlate and phenomenon from mere coincidence."
—Jordan Ellenberg, author of *How Not to Be Wrong:*
The Power of Mathematical Thinking

"An exciting addition to the ranks of books exploring the mysteries of chance and coincidence in the vein of *The Black Swan* and *The Improbability Principle*."

—David J. Hand, professor of mathematics at Imperial College and author of *The Improbability Principle*

"*Fluke* is going to surprise you as a lay reader. You will learn, for instance, that DNA matching is not always clear evidence of guilt in court, and that meeting an acquaintance on your only trip to Disneyland may not be as unlikely as you thought. Mathematician Joe Mazur has written an excellent book, where he shows that probabilistic outcomes are often non-intuitive and unexpected."

—Florin Diacu, author of the award-winning book *Megadisasters: The Science of Predicting the Next Catastrophe*

FLUKE

ALSO BY JOSEPH MAZUR

Euclid in the Rainforest:
Discovering Universal Truth in Logic and Math
(2006)

Zeno's Paradox: Unraveling the Ancient Mystery
Behind the Science of Space and Time
(2008)

What's Luck Got to Do with It? The History, Mathematics,
and Psychology of the Gambler's Illusion
(2010)

Enlightening Symbols: A Short History of
Mathematical Notation and Its Hidden Powers
(2014)

EDITED BY JOSEPH MAZUR

Number: The Language of Science
(2007)

FLUKE

The Math and Myth of Coincidence

JOSEPH MAZUR

BASIC BOOKS

A Member of the Perseus Books Group

New York

Published by Basic Books,
A Member of the Perseus Books Group

Books published by Basic Books are available at special discounts for bulk purchases in the United States by corporations, institutions, and other organizations. For more information, please contact the Special Markets Department at the Perseus Books Group, 2300 Chestnut Street, Suite 200, Philadelphia, PA 19103, or call (800) 810-4145, ext. 5000, or e-mail special.markets@perseusbooks.com.

Library of Congress Cataloging-in-Publication Data

Names: Mazur, Joseph.
Title: Fluke : the math and myth of coincidence / Joseph Mazur.
Description: New York : Basic Books, [2016] | Includes bibliographical references and index.
Identifiers: LCCN 2015043288| ISBN 9780465060955 (hardcover) | ISBN 9780465040001 (ebook)
Subjects: LCSH: Coincidence theory (Mathematics) | Simultaneity (Physics) | Coincidence. | Chance.
Classification: LCC QA612.24 .M39 2016 | DDC 514/.2—dc23 LC record available at http://lccn.loc.gov/2015043288

10 9 8 7 6 5 4 3 2 1

To my inspiring daughters,
Catherine and Tamina

Contents

PART 4: THE HEAD-SCRATCHERS

Introduction

My uncle Herman once packed a year's course on metaphysics into a single short sentence: *Everything that happens just happens because everything in the world just happened.* He was lecturing me at a very impressionable time in my life, when my other uncles, his younger brothers, were teaching me how to read a racing form, hoping to bring me into the family gambling pastime. I was just ten years old at the time, and didn't know what to make of Herman's baffling adage. For years it was a recording in my brain, incubating there, replaying, until sometime in early adulthood when its meaning finally hatched. As a child, I was always questioning why certain things happened and others didn't, and as for most children, the answers came through repeated circulations of *what-if*s.

Herman's younger brother Jack was knocked unconscious at a high school boxing match, and for the rest of his life he suffered from headaches and some kind of mental health disorder judged serious enough for him to be committed to life at an asylum. He got weekly shock treatments at Greystone Park, a place that at once upon a time was officially called the New Jersey State Lunatic Asylum. Even the name of his treatment was shocking—electroconvulsive therapy. For half of Jack's life he endured the agony of those brutal jolts going through metal plates tightly sandwiching his head. One can only imagine what that horrid experience was like, a torture like no other, worse, he said, "than being continually stung by a million hornets." Each shock lasted no more than a nanosecond, but its beastly memory dawdled to return shuddering aftershocks.

Except for the stubble of gray whiskers covering his pock-marked cheeks, Jack never came across to me as strange. He had the best jokes, the warmest genuine smile, and the best adventure stories—told as if they actually happened.

So, my ten-year-old mind spent some time pondering the what-ifs of timing, as if the knockout was the real cause of all of Jack's abnormality, and as if I could turn back the clock to have one of my favorite uncles live a normal life. What if he had been sick that day and hadn't gone to school? What if his boxing opponent had been sick that day, or . . . what if Jack had knocked out the other guy first? Two special events coincided at one moment of time. They always do, of course. But the knockout was the result of a direct hit to the head at the precise moment when Jack's guard was just too low to protect him. Too low, and too slow.

My childhood was filled with what-ifs in hopes of amending unpleasant timings, but the most poignant one happened shortly before my thirteenth birthday. I was on my way home from school, riding my red Raleigh three-speed bike along a flat cracked concrete sidewalk, when a stone hit the spokes of the front wheel and ricocheted to the door of a parked car. I braked and turned to see who had thrown it. At that moment the world suddenly turned red. I could still see. It was as if my stunned brain had not yet processed what had happened. Through the blood rushing from my eyelid I could see a boy across the street ready to throw his next stone. He didn't seem to understand that he had already hit me in the eye. I screamed and fell to the pavement, not fully grasping what had just happened. The next moment I remember is sitting up in a hospital bed with my left eye bandaged, learning that I would probably never again see from that eye. Those what-ifs were so strong that it took years for them to calm down. When I put the quandary to my mother, she gave me the consolation that I was lucky that the rock had not hit my skull and dumbed my brain.

"Could it really have dumbed my brain?" I asked, as if my mother knew something about neuroscience.

"Yes, of course," she answered. And I took that as clinical certainty.

But my mother's consolation didn't stop the what-ifs from doomed efforts to bring back the sight of my left eye. *What if the rock's trajectory had veered off by one degree? What if I hadn't stopped to look around? What if that first rock hadn't hit the spokes?* A few years went by before I learned that luckless coincidences are the battle scars of life. Like the wrinkles of an old face, they are the high scores of an active life, the markers of the roads taken. Life itself is an endless sequence of flukes and coincidences leading to some successes, some failures, some embarrassments, and some pleasures. We shall never really know the fortune and misfortune milestones along our roads-not-taken. Our decisions at forks and crossroads in a tangle of flukes and coincidences settle our destinies, to maximize our pleasures and minimize our failures in the face of all that life throws at us.

Coincidences make magnificent stories. We take them as surprising events, marvel at their rarities, and ignore any sensible explanations, even though many of the finest can be explained as mathematically predictable. Tell a coincidence story at any social event and you have all ears at attention. Why? Because, in this enigmatic galaxy, it transmits a strong sense of inclusive human connectivity, encourages evidence of existential significance, and validates our longing for individuality.

This book is a collection of bewildering encounters and phantasmagorical stories that remind us of how enormous and how small the world actually is. It includes practical mathematical methods for appraising a story's likeliness and examines the nature of coincidence frequency to explain why coincidences deceptively surprise us when they happen. It tours the early development of mathematical tools for understanding randomness, subsequently leading us to think about coincidences as consequences of living in an enormous world of a large number of random possibilities.

There are two classic problems that give us mathematically proper ways to gauge coincidences. One is a counterintuitive poser: the birthday problem, which tells us that in any group of twenty-three people the odds are better than even that two people in the group will have the same birthday. The other is the monkey problem, which asks: if given a large enough amount of time, could a monkey, randomly hitting the keys of computer keyboard, write the first line of a Shakespeare sonnet? These two problems, along with the law of large numbers, the theory of hidden variables, and the law of truly large numbers, give us a reasonable understanding of why coincidences happen far more frequently than we expect. This last law, the law of truly large numbers, is a philosophical adage, and the central argument of the book. In a nutshell it tells us that if there is any likelihood that something could happen, no matter how small, it's bound to happen at some time. It's not a theorem that can be proven. After all, I used the phrase "bound to happen," which is as ambiguous as any phrase can be. But it gives a sense of just how common coincidences are.

The book is in four parts. Part 1 presents a small group of coincidence stories to ponder before attempting to understand the frequencies of coincidental events. Each story represents a whole class of stories that have similar analytical features. Part 2 covers all the mathematics you would need to know to be able to understand the central argument of the book. In Part 3 we return to the ten representative stories of Part 1 to analyze their frequencies to find that absolute randomness, as a theory, is not the same as absolute randomness in the real, physical world. Part 4 gives us a fun chance to explore those coincidences that defy analysis, such as odd and tragic stories of incriminating DNA evidence, lucky scientific breakthroughs, rogue market bets, ESP wonders, and coincidence plots in fiction and folklore. Each chapter in Part 4 is fairly independent of the others.

When you reach the end of this book, you will be looking at the mysteries of coincidences through a curious lens focused on how they happen, and how they preserve their wonders. This book will not only reveal the surprises behind those frequencies to explain why coincidences happen, it will also change the way we see things. Most of our daily events or circumstances don't come to us in simple ways, but are connected to so many other events and circumstances that are beyond our notice. Any single event is a result of many others, along with complex concepts beyond our reach. So, though I will use mathematics to explain why some coincidences happen, I will also accept—and, at times, argue for—some notions of fate, when rational explanations seem weak, and concede that it sometimes feels good to believe that there is a grand plan governing that which we cannot explain.

Although I do admit to crushing the impression that coincidences are rare, it is never my intention to quash the mystiques and charms of a good story. If I shatter the aura of anyone's coincidence encounter, it is done only to appraise it from a mathematician's viewpoint. I have no wish to thwart the seeds of great storytelling. You could argue with me over the question of destiny or fluke, and might even convince me that no one knows enough about the universe to say definitively whether coincidences are mysteriously fated by some deeply significant design. I might even agree with you that flukes, by definition, have no rational explanation for happening. But the mathematics is real and clear. Coincidences happen far more often than we think, mostly because we live in a larger-than-imagined world with over 7 billion people making decisions every second leading to an unimaginably vast number of dependent outcomes. It gives us a causality universe that is unimaginably vast and complex, a place where unlikely events happen simply because there are so many possibilities and so many of us available to experience those possibilities. Things coincide by mere chance alone, without any apparent presence of cause, although a*pparent* is one of those tricky words whose meaning is hard to pin down.

We all have our personal stories of coincidence. Mine inflates to coincidence status, simply because it takes on an importance to me. Meeting my wife on Moratorium Day in 1969 in a crowd of a hundred thousand people on the Boston Common seems astonishing to me because it was momentous in the decisive path of my life. Such events in life compel us to wonder about the what-ifs of decisive timings: what if I had stopped to tie my shoelaces during the march to the Common while two hundred marchers passed ahead of me, or what if I had entered the Common just 10 yards to the north? But is that really a coincidence, or just a happening told in hindsight?

So far, I have used the word *coincidence* twenty-four times in this introduction as an acceptable synonym for "a chance happening," or more narrowly, for a convergence of characters or objects in time and space. Until now, I have assumed its meaning to be self-explanatory, but to be more precise, let's agree on this more formal definition.

> **coincidence** \ n–s: A surprising concurrence of events or circumstances appropriate to one another or having significance in relation to one another but between which there is no apparent causal connection.[1]

Somehow, the colloquial use of that word tends to drift to an interpretation that ignores the part that requires surprise and also expects any cause to be nonapparent. For us here, we maintain that any coincidence must have a presence of surprise and, if there is any cause at all, it must be one that is nonapparent.

A coincidence's surprise is closely tied to its nonapparentness of cause. When we use the phrase *nonapparent cause*, we simply mean that there is a cause that is unknown to the general public. Coincidences do have causes. So, yes, the question of relativity emerges: to whom is it unknown? For our purposes, we will assume that by *general public* we mean the person who experiences the coincidence as well as anyone who is told the story.

Fluke, on the other hand, has a similar meaning without the caveats of surprise and apparent cause.

> **fluke** \ n-s [origin unknown]: an accidental advantage or result of an action: an extraordinary stroke of good or bad luck.[2]

And *serendipity* is restricted to positive events.

> **serendipity** \ n: The occurrence and development of events by chance in a happy or beneficial way.

Almost all stories are told through a series of events—meetings of characters and objects—happening in some time. Oedipus kills a man on the road to Thebes and thus through a chain of events he sleeps with his own mother. What's the apparent cause there? It's the chain, each link with an apparent cause. It's worth noting that every coincidence is a chain of events with each link causal, even in the real nonfiction world.

Neil Forsyth, the literary essayist and *professeur honoraire* at the University of Lausanne, calls the coincidence chains "the delight in the unexpected."[3] He is referring to fictional coincidences in Dickens, but that delight in the unexpected is true in the real nonfiction world as well. It comes from a deep need and strong desire to make sense of the odd unfamiliar, a need that was once-upon-a-time fundamentally vital to human understanding and protection from the unknown.

For many of the most surprising coincidences, their nonapparent causes might be too deep to ever be discovered. It's easier to believe they are unexpected than in the fact that extraordinary stuff just happens; it's more comforting and promising to our own prospects. In any case, they do amuse us.

The sum $1^3 + 5^3 + 3^3$ happens to be 153. Is that a coincidence? The cause is not apparent. It is possible that there is no cause at all. Or consider this thoroughly random sequence of sixty digits:

458391843333834534555555555555
185803245032174022234935499238

We might be suspicious of that continuous string of 5's in the middle. Those 5's might be "hot," but the mathematics tells us to be not so surprised. It even predicts that such a sequence of equal digits will happen far more regularly than we think.

Coincidences are omnipresent. It all comes down to noticing. Just before writing this introduction, I was vacuuming too close to my 2,262-page dictionary. As always, to protect its thick binding, it was open to a page a bit beyond the middle. The vacuum head suddenly sucked in a whole page. Consoling myself, I thought, "Will I ever really need page 2072? Not likely." Less than an hour went by before I went to look up the exact wording of the entry for *serendipity*. You can guess what page that word was on. When you write a book on coincidences, you notice more than ever.

PART 1

The Stories

Coincidence

It starts as true story,
first wondrous, and rare,
then colossal collisions
of galactic affair
strikes with so much surprise,
we're confused between thoughts
that it might be just fluke
we believe in a lot,
and what if it's not.

—J. M.

LIFE IS FILLED with expectancies, to-dos, and benign pleasures, but bewildering encounters and phantasmagorical stories give the blisses of being alive. Here is a browsing of just a few accounts of how our world is both enormous and small, and of how we come to distinguish flukes from coincidences. We shall return to these stories in Part 3, after we have some machinery to illuminate their hidden quantitative elements.

Chapter 1

Exceptional Moments

REMEMBER THAT TIME when you were leisurely strolling along a street in a foreign city, Paris or Mumbai perhaps, and bumped into an old friend you had not seen in a long time? That old friend that you bumped into: what was he doing there in your place and time? Or remember that moment when you wished for something and it happened just as you wished? Or the fluke of hard luck you had when everything went wrong during your vacation because of unfortunate timing? Or that time when you were astonished to meet someone who shared your birthdate? They were times when you must have had a sudden feeling of synchronicity that shrank the universe, an illuminating transformation that magnified your place in the cosmos. You felt part of an enlarged and centered circle of humanity with just a few persons—or perhaps just you—at that center.

Did you ever pick up a phone to call someone you hadn't called in a year and, before dialing, hear the person on the line? It happened to me in 1969. Thinking about it, it seems more likely to happen than not. After all, a whole year had gone by—365 days when it didn't happen. Add to that the number of days the year before, another year when it didn't happen. And add to that the number of days from then to now. It never happened again.

Now we are talking about a serious amount of time when the coincidence did not happen.

Imagine this story. You are sitting in a café in Agios Nikolaos on the island of Crete, when you hear a familiar laugh at a table in a nearby café. You turn to look at the person, a man. You cannot believe that he is your own brother. But there he is, unmistakably your own brother. He turns toward you and he is as surprised as you. It happened to me in 1968. Neither one of us knew that the other was not back home in New York or Boston.

Or imagine this. You are browsing used books in a bookstore far from home when you come across a book you remember from your childhood. You open it and find your own inscription. It is a copy of *Moby Dick* with your own name on the inside cover and your own marginal markings throughout the book. It was a book you had in college. It happened to a friend who told me that he was browsing the shelves of a used bookstore in Dubuque, Iowa, a city he had never been to before.[1]

In 1976 my wife, my two children, and I had been touring through Scotland, when on one snowy day our Vauxhall car broke down in the small town of Penicuik. A mechanic at the only garage in town told us that the problem was our alternator, and that it would be three days before he could replace it. We headed to the nearest pub, hoping to spend the night. The publican was a man of few words, but when we told him that we were from America, he livened up to proudly say, "Next week we will have someone from America coming to sing. You probably know her. I don't know her name, but there is a poster downstairs." He brought us to a large poster announcing a *stovies* night[2] concert by Margaret MacArthur.

"Margaret MacArthur!" my wife and I exclaimed simultaneously. "She is our neighbor. We know her very well!"

The publican nodded, and with blank countenance muttered, "Thought you would."

America is truly a small country.

There are moments when we are struck by magnificent coincidences. They are the foci of nature's web of associations because, especially in the solitude of this digital age, we want to fit into the intimidating world with a sense of self, an identity, a purpose, and a feeling that some parts of our lives have destinies. Daunted by the chilling vastness of the forever-expanding universe in an endless space and time, it is reassuring to know that we are more connected than we think, or that the universe lines up for us.

With any coincidence story is the question of whether there is something in the universe that perturbed time and place enough to trigger the coincidence and conceal its cause. Some people have questioned whether there are metaphysical connections. Some say that there is oneness in this universe, an energy that we cannot be aware of, a force that changes our patterns of behavior, a *knowing* something that we do not know.

Causality is the Western way of interpreting the meaning of events. Nineteenth-century Western causality had a strict classical physics view that the laws of nature govern the movement and interaction of all observable objects. If the variables of the present state are precisely known, then the future is completely predictable. In other words, any predictions of the future are tied to whatever we can know of the past and present. However, by the early twentieth century, with the invention of quantum mechanics, Western philosophy took a radical shift of viewpoint: observable objects are driven by non-observable events of the quantum world, governed by simple, wondrous rules. One such rule claims there are no roads not taken. Every particle is ordered to follow not just one path, but also every possible path with a probability that depends on the path. Predictability, in that quantum mechanics point of view, is limited to probabilities that an object will be somewhere on each path and in a particular state. In other words, careful observation of exactly what happened in the past only gives us uncertain probabilities of what might happen in the future.

Of course, there is always the question of what causes one person to choose a path forward. We are not talking about the mechanical path of an object. Why did you, dear reader, choose to read this far into this book? You have free will that has almost nothing to do with classical physics, or the path of observable objects, or the new physics. The coincidences of this book are related to decisions that people make, roads taken and those not. Human decisions are a matter of free will, where neither relativity nor quantum mechanics come into play, although there are always other strong external influences. We decide on a path. Someone else decides on a path. Then, bam! The paths meet, and we have no apparent cause. The problem with apparentness is that it requires an observable object traveling on an observable path. So, unless there are brain-wave connections between distinct individuals, free will trumps all quantum influences.

There is also the Eastern way, however. The Chinese, for instance, have their Tao, in which opposites cancel out to make the whole and total picture. In it the nothingness is also part of the whole. A block of stone can become a sculpture defined by the remaining stone and the stone that has been carved away. It is surely a different way of thinking. And still the Tao belief is certainly different than any theology that looks at the world as if everything in the world, from cells of organisms to subatomic particles of minerals, is prearranged from the time of creation, and laws governing causality can be broken only if ratified by God's will. The Taoist believes that coincidences are in the sympathy of all things, and for that reason all events in the world stand in one relationship beyond any causality and any appearances. In other words, there are no flukes. But that same Taoist also believes that underneath there is a hidden rationality. The revered *Tao Te Ching*, now some 2,500 years old, tells us:

> Heaven's net is wonderfully vast and enveloping;
> Though wide-meshed, nothing slips through.[3]

Just as all parts of a whole work in harmony to complement one another, so, too, all events in the world stand in one meaningful relationship with the whole, which is in central "meaningful" control.

Walt Whitman also told us that we have some connection to the All, and that there is a moral purpose and intention that we are forced to follow unconsciously. He put it this way:

> As within the purposes of the Cosmos, and vivifying all meteorology, and all the congeries of the mineral, vegetable and animal worlds—all the physical growth and development of man, and all the history of the race in politics, religions, wars, &c., there is a moral purpose, a visible or invisible intention, certainly underlying all . . . something that fully satisfies . . . That something is the All, and the idea of All, with the accompanying idea of eternity, and of itself, the soul, buoyant, indestructible, sailing space forever, visiting every region, as a ship the sea.[4]

Chapter 2

The Girl from Petrovka and Other Benign Coincidences

What connection can there have been between many people in the innumerable histories of this world, who, from opposite sides of great gulfs, have, nevertheless, been very curiously brought together!

CHARLES DICKENS, *Bleak House*[1]

IF YOU LEAVE your house, a great many encounters and happenings are possible. The probability of each may be small, but when we group them together and ask for the probability that at least one of them will happen, the likelihood goes up. These stories are just ten of many that effectively represent ten characteristic classes. They will be analyzed in Part 3.

Story 1: The Girl from Petrovka

Class: Lost, unlikely to be found, objects accidentally found by someone deliberately looking for them

One of the most celebrated coincidence stories involves actor Anthony Hopkins. After being cast to play the part of Kostya in a movie version of *The Girl from Petrovka*, Hopkins spent some time searching for the novel in bookstores near London's Leicester Square Underground station. Unsuccessful in his search and about to return home, he noticed a book lying on one of the benches of the same Underground stop. It was not just a copy of *The Girl from Petrovka*, but the lost copy belonging to its author, George Feifer.

It is a truly bewildered story. I should be forced to concede that this one would be so foreign to any reasonable theory of the frequency of coincidences that I should have to congratulate the story for escaping any explanation. But in truth it does not escape analysis. George Feifer told me the true story himself: He had used a copy of the American edition of *The Girl from Petrovka* to highlight words that required British translations for the UK publication of his book. He submitted his translations to the British publisher and checked them on the copy plates. One day he met a friend in Hyde Park Square and gave the friend his marked-up American edition. In a daze of the moment, the friend put the book on top of his car and, late for an afternoon meeting with a girl, speedily drove off. On seeing Feifer on the movie set, Hopkins told Feifer that he found the book at an Underground stop. I wrote to Hopkins for his side of the story. Predictably, he never replied.

Story 2: Jack Frost and Other Stories

Class: Unexpectedly found familiar personal objects not searched for

A comparable story involves writer Anne Parrish. According to the original telling (a story very different from the many floating through cyberspace), while in Paris, after attending Mass at Notre-Dame and visiting the bird market on a sunny June Sunday in 1929, Anne and her husband, industrialist Charles Albert Corliss, stopped at Les Deux Magots for lunch. Leaving Charles alone with his wine, she strolled the bookstalls along the walls of the Seine. It was not unusual for her to spend hours rummaging rows of books on long tables. On that day, she found Helen Wood's *Jack Frost and Other Stories*. After a short haggle with the bookseller, she paid one franc, ran to her husband who was still sitting with his wine, excitedly put the book in his hands, and told him that it was one of her favorite books when she was a child. He slowly turned the pages. After a few moments of silence, he handed the book back to her opened at the flyleaf where penciled "in an ungainly childish scrawl, was: 'Anne Parrish, 209 North Weber Street, Colorado Springs, Colorado.'"[2] It had been her book when she was a child.[3]

Story 3: The Rocking Chair

Class: Requiring reasonably precise time and space and not human chance meetings

A coincidence has to be more than a story forced to give a surprise or to hide its cause. Here's one that happened to me some years ago. My wife was pregnant and her aunt told her that she must have a comfortable rocking chair for nursing the newborn baby. She sent a check to cover the purchase of a new rocking chair. My brother had the perfect rocking chair, and my wife and I found that same chair at a furniture store in Cambridge. It was uncommonly wide, a Shaker design with thin black spindles and a high back. But the chair was not in stock, and so we asked that it be delivered to my brother's house in Cambridge when it was available for delivery. We would pick it up from there and bring it home to Vermont at our next visit. Some weeks later my brother and his wife were hosting a small gathering. Someone sat on their rocking chair, and the entire chair collapsed underneath him, breaking to pieces. Embarrassed, my brother courteously told his guest not to worry. At that precise moment the doorbell rang and our new chair was delivered. One can only imagine the surprise at the party, when my brother took the golden opportunity to console the guest by saying, "Oh, that's fine; we just called for a replacement."

Story 4: The Golden Scarab

Class: Dream coincidences in somewhat generous time and space

A young woman patient was telling Swiss psychiatrist Carl Jung of her dream about a golden scarab. We have Jung's version: "While she was telling me of this dream I sat with my back to the closed window. Suddenly I heard a noise behind me, like a gentle tapping. I turned round and saw a flying insect knocking against the windowpane from outside. I opened the window and caught the creature in the air as it few in. It was the nearest analogy to a golden scarab that one finds in our latitudes, a scarabaeid beetle."[4] Jung goes on to say, "We often dream about people from whom we receive a letter by the next post. I have ascertained on several occasions that at the moment when the dream occurred the letter was already lying in the post-office of the addressee."[5]

Story 5: Francesco and Manuela

Class: Chance meetings of humans in precise timing and space

My wife and I were in a van riding along the hairpin road that cut through la Costa Smeralda, the eastern coastal hills of Sardinia, high above the clear emerald waters of the Tyrrhenian Sea. They were breath-halting moments when our Italian driver was gesticulatively pointing out historical sites while turning his head back and forth for quick peeps of the dangerous curves ahead and glances at passengers in the backseats. We were spending some time at Studitalia, an Italian language school in Olbia, a picturesque small port city on the northeast coast of Sardinia. It was a weekend. And—as it happened every weekend—the school had offered its students an excursion immersed in Sardinian culture and beauty. The driver was Francesco Marras, the school's director.

A student sitting in the front passenger seat asked him when and how the school started.

"Well—" he began his answer, thinking about the story he was about to tell as the van wigwagged from one side of the road to the other seconds before the next curve. "When the school opened just three years ago in 2010 there was only one student." In a typical Italian way, he used his right hand to illustrate his story, and his left to blithely steer the van.

And so, we learned how on that opening day Francesco went to the lobby of the Hotel de Plam to meet the school's first student, Manuela from Madrid, for an orientation excursion, which involved a boat trip to the magnificent Isola Tavolara, an enormous flat-topped rock island some 3 miles offshore. Since Francesco and Manuela were early and their boat was late, they went off to a café for a drink. They sat for an hour, chatting in Italian. Manuela talked about where she lived in Spain, about her work, her boyfriend, and her interests. Francesco talked about the school. Soon Francesco began to wonder why Manuela would want to take Italian when her command of the language was quite excellent.[6] When he finally asked about the level of Italian she expected to learn at the school, the mix-up became clear.

"Learn Italian? Why, do you think I need Italian lessons?" she asked.

The confusion lasted several more minutes before Francesco realized that Manuela was the wrong Manuela, who had expected to meet in the hotel lobby someone by the name of Francesco!

They both returned to the hotel lobby to find the other Francesco interviewing the other Manuela for a job she neither expected nor wanted.

Why is this story so surprising? Because it has been humanized as a story with a place and time, specific names, a colorful character who seems to be telling the truth. Intellectually, we are not fooled. We know that with large numbers of possibilities these encounters happen, and that they are not so unusual.

Story 6: Albino Taxi Driver

Class: Chance meetings of humans under generous timing and generous space

These stories and others like them are more common than we generally think. We are told such stories, and many of us have experienced them. Just the other day I met a woman who told me a wonderful story: In Chicago one day she got into a taxi driven by a man with albinism. Three years later she got into the same man's taxi in Miami. "Now, what are the chances of that?" she asked me. Yes, this is a wonderful story, but let us deconstruct it. Taxis frequent particular neighborhoods. The woman is an executive of a private equity firm, someone who often takes taxis in different major cities. Taxi drivers that do not have albinism are not as distinguishable; so, a person who uses taxis often, might expect to hail a taxi without noticing that the driver is familiar, unless that driver happened to be a person with albinism. Still, I would agree that we should indulge in some sort of fascination with the fact that Miami and Chicago are 1,200 miles apart.

Story 7: Plum Pudding

Class: Associations with familiar objects

And here is another story about a doorbell ringing to announce a coincidentally surprise visitor. I learned this one along with several others in *L'Inconnu: The Unknown* by early twentieth-century astronomer Nicolas Camille Flammarion.[7] It's one of those double coincidences, the kind that brings some astonishment, and then a new surprise happens to top it, turning it into a triple coincidence.

Flammarion writes that Emile Deschamps, a celebrated nineteenth-century poet, told this tale. Deschamps was a young boy at boarding school in Orléans, France, when he met an English émigré by the curiously non-English name of M. de Fortgibu. Dining at the same table, M. de Fortgibu suggested that the young Deschamps taste a dish that was almost unheard of in France, plum pudding.

For ten years Deschamps, having not seen or heard of it again, forgot about his discovery of a plum pudding that oddly contained no plums. Ten years later, passing a restaurant on the boulevard Poissonière that displayed the strange pudding on its menu, Deschamps was reminded of M. de Fortgibu. He ordered a slice, but was told by the counter ladies that a certain gentleman had ordered the whole pudding. One of the women turned toward a man in a colonel's uniform who was eating at one of the tables.

"M. de Fortgibu," she called out, "would you have the goodness to share your plum pudding with this gentleman?"

Deschamps didn't recognize M. de Fortgibu.

"Of course," M. de Fortgibu answered. "It would give me great pleasure to share a part of this pudding with the gentleman."

It wasn't likely that he recognized Deschamps, either.

Now, that should have been the full coincidence, but it was not. Some years passed. Deschamps had not seen or thought about plum pudding. And then one day he was invited to a dinner at the home of a lady who announced that an unusual dish was to be served: a real English plum pudding.

"I expect a M. de Fortgibu will be there," he joked.

The evening of the dinner arrived. A magnificent plum pudding was served to the ten guests seated while Deschamps told the coincidence story of M. de Fortgibu and the plum pudding. Just as Deschamps completed his story, everyone heard the doorbell ring and M. de Fortgibu was announced.

You and I would think this was all planned. Deschamps thought so. Perhaps the dinner host had used Deschamps's little joke to build a joke of her own. But no! It gets even more interesting. By this time M. de Fortgibu was an old man who walked with a cane. He walked slowly around the table looking, for someone in particular. As he came close, Deschamps recognized him. Surely it was he.

"My hair stood up on my head," Deschamps said on relaying this story sometime later. "Don Juan, in the *chef d'oeuvre* of Mozart, was not more terrified by his guest of stone."

But Deschamps was not the person the newcomer sought, It turned out that M. de Fortgibu (the same) was also invited to dinner, but not *that* dinner. He had mistaken the address and rung the wrong doorbell. It was a triple coincidence that must be so rare that you would think that the chances of its happening in one's lifetime must be staggeringly close to zero. Yet it did happen, if we can trust M. Flammarion.[8]

"Three times in my life have I eaten plum pudding," Deschamps reflected on his confounded experience, "and three times have I seen M. de Fortgibu! A fourth time I should feel capable of anything . . . or capable of nothing!"

Flammarion, the respected astronomer who has lunar craters, Martian craters, and asteroids honoring his name, was a collector of coincidences. As he was known for being a collector, people sent him their stories. He collected hundreds. Some quite astounding! Many were sent to him anonymously from all parts of the world, so it is very hard to trust their truthfulness, even though he does say that some had multiple witnesses, that some had a sincerity he vouches for, and that others have "all the marks of good faith."

Story 8: The Windblown Manuscript

Class: Coincidences dictated by natural causes

The most remarkable coincidences are those of Flammarion's personal experience. One is a captivating story that suggests that there are some miraculous forces looking out for us, chance perhaps, or unknown forces that parallel those of nature. He was writing his eight-hundred-page popular treatise on the atmosphere.[9] It was to become his definitive work. At the end of the nineteenth century it was very celebrated both for its detail and its accessibility. Just at the point where he was busy writing the third chapter of the fourth section, a chapter about the force of wind, the most extraordinary thing happened. It was a cloudy midsummer day. He was in his study. One window, facing east and overlooking some chestnut trees and avenue de l'Observatoire, was open. Another window faced southeast with a magnificent view of the Paris Observatory, and a third faced south onto rue Cassini. He had just finished writing: "*Les vents de nos climats, qui nous paraissent si capricieux et si variables, vont nous laisser apercevoir derrière eux les règles auxquelles ils obéissent*" (The winds of our climates, which appear so capricious and variable to us, will let us see behind them the rules which they obey).[10] A sudden southwest gale blew open the window overlooking the observatory, lifted the leaves of his manuscript—a whole chapter—from Flammarion's desk, and carried them off to the street below. Some fell among the trees and some scattered toward the observatory. Worse yet, a soaking downpour followed. That was the first coincidence of that day.

Flammarion realized that it would have been useless to go searching for all his missing pages. He wrote, "To go down and hunt for my pages would seem to me to be time lost, and I was very sorry to lose them."[11] What happened next was truly astonishing. A few days had passed when a porter from Librairie Hachette, the publisher of Flammarion's books, a mile from his apartment, brought him all the missing leaves.

Story 9: Abe Lincoln's Dreams

Class: Dreams that come true

Abraham Lincoln told this prophetic dream to his wife, Mary Todd, one night at dinner, shortly before he was assassinated.[12]

"About ten days ago I retired very late. I had been up waiting for important dispatches from the front. I could not have been long in bed when I fell into a slumber, for I was weary. I soon began to dream." Lincoln then went on to say that in his dream he had left his bed to go downstairs. He may have actually done so.[13] Downstairs—presumably in the White House—he heard a group of mourners sobbing. From room to room he searched for the mourners, and though the rooms were lit, he couldn't see any. Yet the sounds were all around, as if the mourners were invisible in each room. Even though it was an alarming dream, he wondered about the meaning of it. When he came to the East Room, a corpse in funeral garb was resting on a catafalque, with several soldiers standing guard. Mourners were standing all around, weeping. The corpse's face was covered. "Who died in the White House?" he asked one of the soldiers. "The president," answered the soldier. "He was killed by an assassin!"

The crowd then began to wail so loudly that Lincoln awoke from his dream. He said that he did not sleep more that night and that he had been haunted by that dream ever since he had it.

"That is horrid!" Mary said. "I wish you had not told it. I am glad I don't believe in dreams, or I should be in terror from this time forth."

"Well," said Lincoln in a very somber voice and with a glum face, "it is only a dream, Mary. Let us say no more about it, and try to forget it."

He had other premonitory dreams before almost every event of the war. There were repeated omens of a Union victory: One happened the night before the Antietam victory, and another a few nights before Gettysburg. There were others preceding Sumter, Bull Run, Vicksburg, and Wilmington. One happened on April 13, 1865, the night before he was shot at Ford's Theatre. It was a very vivid one. During the day of April 14, General Grant informed the cabinet that he was awaiting General Johnston's surrender. Lincoln, in his deep and confident voice, then said, "We shall hear very soon, and the news will be important." When Grant asked him why he thought so, Lincoln said, "Because I had a dream last night; and ever since this war began I have had the same dream just before every event of great national importance. It portends some important event that will happen very soon."

It seems that all the dreams he talked about were prophetic. Johnston surrendered to General Sherman on April 26. The war was finally over. And the man who dreamed these dreams was no longer alive. Three days after Lincoln's assassination, Gideon Welles, the secretary of the navy, a man who was present at Lincoln's last cabinet meeting, wrote these words in his diary:[14]

> Great events did, indeed, follow, for within a few hours the good and gentle, as well as truly great, man who narrated his dream closed forever his earthly career.

Lincoln's last cabinet meeting was called for 11:00 a.m. on Good Friday, April 14. Frederick Seward, the assistant secretary of state, was at that meeting. He wrote about the meeting in *Leslie's Weekly*, a newspaper illustrated by wood engravings and daguerreotype photography:

> The conversation turning upon the subject of sleep, Mr. Lincoln remarked that a peculiar dream of the previous night was one that had occurred several times in his life,—a vague sense of floating—floating away on some vast and indistinct expanse, toward an unknown shore. The dream itself was not so strange as the coincidence that each of its previous recurrences had been followed by some important event or disaster, which he mentioned.

The usual comments were made by his auditors. One thought it was merely a matter of coincidences. Another laughingly remarked, "At any rate it cannot presage a victory nor a defeat at this time, for the war is over."

A third suggested: "Perhaps at each of these periods there were possibilities of great change or disaster, and the vague feeling of uncertainty may have led to the dim vision in sleep."

"Perhaps," said Mr. Lincoln, thoughtfully, "perhaps that is the explanation."[15]

Story 10: Joan Ginther

Class: Good and bad gambling luck

How should we think of the luck of a woman who wins a lottery four times?

July 14, 1993, Joan Ginther walks into a Stop N Shop in Bishop, Texas, buys a few Lotto Texas scratch-off lottery tickets, and wins $5.4 million. That's local news.

The same woman walks into a minimart some years later, buys a few Holiday Millionaire scratch-off tickets, and wins $2 million. Texas news.

Two years go by. She buys a few Millions and Millions from the Times Market on US Highway 77 in Bishop, and wins again! Another $3 million! National news.

Another two years go by. She walks into the same Times Market, buys fifty dollars' worth of Extreme Payout, and wins another $10 million. Now it's international news! "Who is the lucky four-time lottery winner?" John Wetenhall of ABC World News asked one week later.

The odds of that happening to a particular person are 1 in 18 septillion, so unlikely that it should happen to that person only once in a quadrillion years.

Some people believe that Joan Ginther, a retired mathematics professor with a PhD from Stanford, had gamed the system, cheated in some way, or perhaps cracked the lottery's algorithm that determines where winning scratch-off tickets are shipped. Others felt that she had won by clues from displayed numbers that give information about which cards are winners. But many of the people of Bishop, a little farming town of 3,300 inhabitants, believe that "it was Joan's reward from God."

Such multiple wins are rare, yet not surprising to statisticians who know that rare events will happen just by chance: Winning a lottery four times is a rarity when considered as a per-person event, but is a reasonably common event when considering the larger population. In fact, the odds are quite good that wins like Ginther's have a good chance of happening in a population of almost 320 million Americans. Her wins seem striking only because we are viewing them as happening to one specific person, Joan Ginther.

Considering that there are twenty-six major legal lotteries in the United States alone, with ticket sales of over $70 billion by people who play frequently, a four-time win is not only bound to happen, but should happen fairly often over the years.[16]

Chapter 3

Meaningful Coincidences

THERE ARE CONNECTIONS that simply cannot be explained as chance groupings of time and space alone. Those "coincidences" are connected so meaningfully that their chances share a hugely high degree of improbability.

We might ask for a cause, and look for a meaning. Cause and meaning are two different things. The cause of an event is the principal reason that the event happens. There are causes that are not determinable, causes that are too deep for us to catch, and causes that are too vague to understand. A cause could have multiple layers of understanding. A tree falls when a large enough cut is made at its base. On one level, the cut might be the cause of its fall; on another, the cause could be the tree's weakness in balancing itself after the cut; and still on another, the cause could be that the tree's trunk is so rotten that it falls regardless of the cut. Meaning, however, is different.

Here is a point to consider: As you read this sentence, the sun is radiating inside the room you are in. Am I right? For some readers, I would be right. It's a reasonable guess that some people are reading this book on a sunny morning, possibly a Sunday morning. If I had written, "As you read this sentence on a Sunday morning lying on your couch in a room with three windows behind you that need cleaning," I probably would have eliminated a large number of readers. Those of you who are reading on your way home from work in, say, the #2 train heading to Flatbush Avenue in Brooklyn, New York, would understand that I am not addressing you—although, coincidentally, I just *did*.

If it *is* Sunday morning and you *are really* lying on your couch in a room with three dirty windows, you might think of the sentence as some spooky coincidence. You might even think you are the only reader. But really, I just made something happen by guessing how many people will be reading this book on a couch and getting this far on a sunny day.

I didn't name the reader. I could have written: "Larry Smith, as you read this sentence the sun is radiating inside the room you are in." The chances would have been slim that some Larry Smith was reading these paragraphs on a sunny day, but not zero.

But that is not what we really mean by coincidence. Any cause would have to involve my presumption that there are (as I can only hope) a substantial number of readers to make the concurrence possible. Would it be a coincidence? No! The cause is evident, and besides, the meaning is marginal. I constructed the sentence to force the possibility. Effectively, I caused it to happen by concocting an image of likely readers in their most probable surroundings. I picked a large city along with a common reading site. The cause was I.

Of course, my concocted concurrence has some meaning, just as any incident does, but not a serious meaning, the kind that touches the psyche, alters body chemistry and stirs a mood to contract muscles, shakes some emotions that constrict or enlarge some blood vessels in the brain. For a coincidence to have significant meaning it must communicate an emotional state, perhaps one that refers to an archetype packed in the history of one's own experience. Our collective knowledge and experience shape our expectations, those anticipations that shape our surprises, the critical feature of any coincidence. My concurrence—if it ever comes true—would not strike one's consciousness with a resounding archetypal connection. It's a fabrication that addresses just a few readers caught in the small range of farfetched possibilities. The meaning of a coincidence is not simply the semantics in the vocabulary of its narration. Every story has linguistic meaning, and some, more than others, have suggestive ideas; however, when we say that a coincidence means something, we expect its story to engage subconscious references that evoke experiences in the depths of one's memory.

I offer the following example of a meaningful concurrence without an apparent cause. Well, maybe not entirely without an apparent cause—you be the judge. On the night of October 19, 2006, my wife's ninety-year-old mother died. A week before, after my mother-in-law announced that she was ready to join her deceased husband, my wife said, "Send me a sign." On October 20, after a heavy rain, the most sharply defined, brilliant, double rainbow appeared, and moments later the two rainbows gradually joined together as one. Was it a coincidence? It could not have happened without the particular timing of my wife looking out the window to notice the event. Rainbows don't last long, and their periods of sharpness are very limited. Was its cause apparent? Well, yes. Scientifically, rainbows are caused by sunlight diffracting through tiny spheres of raindrops in the atmosphere; however, the scientific explanation is not the cause of its timing and being noticed. It may very well have been the promised sign. But what caused the concurrence of timing and being noticed? Whatever it was is not apparent, at least in the sense of how we defined *nonapparent* in the Introduction. It is a case of evident meaning without an apparent cause. It surely touched us, even tingled our spines. For a few moments, that rainbow and its archetypal connection gave meaning to the entire concurrence.

Looking back at the ten representative coincidences of Chapter 2, we find that they all have meaning, but two or three significantly stand out. Story 7, "Plum Pudding," represents the class of stories that highlight associations with familiar objects. Its meaning unfolds over time while a perceptual seed of a single encounter germinates to a maturity of significance in the subconscious. It's a story of reference and association, a story of meeting half-forgotten persons and of experiences, a story of awakened memories and roused conscious connotations of referential incidents. Story 9, "Abe Lincoln's Dream," represents the class of premonition dreams. Lincoln's dream of his own assassination was a subconscious foreboding signal assembled from conscious warnings. It represented an omen of a potential event, the possibility of a crazed behavior of someone who disagreed with a wartime decision. Any president must have assassination anxieties. Although Lincoln's anxieties may have been the cause of his dream, it's the telling of it that has meaning, for it gives the general public a collective sense that leaders, too, have natural anxieties.

It's possible to argue that Story 8, "The Windblown Manuscript," also has significant meaning. Consider its original cause, the connection between a manuscript on the atmosphere and a wind taking away the manuscript. Without that cause there would be no story. But our interest in the story has more to do with the finding of the manuscript than with its topic's connection to the cause of the manuscript's initial disappearance.

Arthur Koestler's *The Case of the Midwife Toad* introduces us to another collector of coincidences, Austrian biologist Paul Kammerer.[1] Kammerer theorized that there are side laws of nature that operate parallel with and independently from the known laws of physical causation. He called them *laws of seriality*, unknown forces that traveled in time and space as waves whose peaks caused us to notice coincidences, both meaningful and meaningless. His story is tragic. Shortly before his suicide in September 1926, this celebrated scientist was accused of falsifying his experiments. The complete scandalous story is long with all sorts of hints that his experiments were sabotaged or perhaps an intractable practical joke gone sour. There are testimonies on both sides of the accusations. But the story for us is about Kammerer's notion of *seriality*. "Seriality," he wrote, "is ubiquitous in life, nature, and cosmos. It is the umbilical cord that connects thought, feeling, science, and art with the womb of the universe that gave birth to them. . . . We thus arrive at the image of a world mosaic or cosmic kaleidoscope, which in spite of constant shufflings and rearrangements, also takes care of bringing like and like together."[2]

His book *Das Gesetz der Serie*[3] is about a wild notion, but one that Carl Jung, Wolfgang Pauli, and Albert Einstein found interesting, at least according to Koestler. It is a strange book, if read from the modern point of view of a twenty-first-century reader who knows something about science. It contains exactly one hundred trivial concordances of events in time and space to use as examples of his theory that coincidences come in bands and series. It's an odd idea, but one that makes you think that it's not so kooky as it first appears and that there is some merit in thinking further. You see, his coincidences were categorically collected, coincidences of things popping up in immediate sequences of events in roughly the same time and place, numbers, pairs of names of unrelated persons, chance meetings of acquaintances, dreams in sequence with real-life experiences, similarity of words materializing in tandem. He was trying to categorically follow the same or similar events that happen at the same time without any apparent causes, so as to build some mathematical or scientific theory. He was collecting empirical evidence in an attempt to figure out whether some unknown laws and principles might be operating behind the scene of space and time that can explain the *seriality*—the frequencies and clusters—of coincidental happenings.

It's reputed that Kammerer would sit on benches in several Vienna parks, taking note of anything happening in the park that could be categorized as a coincidence, say, two people carrying the same briefcase, wearing the same hat, or unexpected meetings. Trivial things like that. Beyond that, he would make notes of numbers of people in the park at different times, how many were women, how many were carrying briefcases, how many were carrying umbrellas. In short, collecting data. Then he would systematically work his data into a quantitative argument to conclude that coincidences are all around us but that we mostly ignore them because we don't expect them. We see them only when we pay attention. And we pay attention mostly when we are told about them, or when they mean something to us. It reminds us of Christopher Chabris and Daniel Simons's famous invisible gorilla experiment that showed the failure to perceive a visible, unexpected object while attention is focused on a task. In that experiment participants were asked to watch a one-minute movie of a basketball game. Players of one team wore black shirts and players of the other team wore white shirts. Subjects were asked to silently count the number of passes made by the white-shirted players while ignoring any passes by black-shirted players. Halfway through the video, a female student in a full-body gorilla suit walked across the court, stopped directly in front of the camera, thumped her chest, and walked off. At the end of the video the subjects were asked whether they had seen anything unusual enter the room. About half of the subjects did not notice the gorilla! A gorilla that walked directly through the center of the court! The gorilla did not contribute to the task; hence, there was a deficiency of attention, and hence, the gorilla was invisible.

And this is part of Kammerer's point. If we are consciously looking for coincidences, we find them all around us. Not just because of our argument that, given enough time with a massive population of events, the most amazing things will happen by sheer chance alone.[4]

I love a good story, and so would not wish to burst the wonder of a surprising incident. But I am also a mathematician who, by professional obligation, must tell the truth. The skeptics will remain skeptical, so the good and astonishing stories will still be told. There is the one about Norman Mailer's novel *Barbary Shore*, a surreal political allegory about a group of six people, each symbolic of an American political view of the time, living in a rooming house in Brooklyn. The principal character is Michael Lovett, an American Marxist-Stalinist. The book appeared in 1951, at the beginning of McCarthyism. A CIA agent read it and arrested Rudolf Ivanovich Abel, a Russian spy in the apartment directly above Mailer's. Mailer had no idea he was living below one of his own characters. That kind of story will always be around, no matter how debunked the coincidence is, partly because it has meaning—subconscious urban concerns of living among unfamiliar neighbors. Tom Bissell in his new book *Magic Hours* tells us that *Moby-Dick* was a flop when it first came out in 1851. Its success as the great American novel didn't happen until 1916, when the persuasive book critic Carl Van Doren accidentally came across a rare and dusty old out-of-print copy in a used bookstore and wrote a glowing essay calling *Moby-Dick* "one of the greatest sea romances in the whole literature of the world." A more recent story is one about Mischa Berlinski's novel *Fieldwork*. It was a sleeper for five years before Stephen King hesitatingly picked it up in a used bookstore and wrote a stunning review of it in *Entertainment Weekly*. It went from minuscule sales to being on the *New York Times* Best Seller list. It was a chance meeting of a book on a shelf in a store that King happened to walk into. These stories have meaning to us through the archetype of hoped success.

Synchronicity

In the early twentieth century Carl Jung introduced the notion of synchronicity as a model for magic and superstition surrounding strange concurrences of events. He did not see coincidences provoked by the unpredictable impressive happenings that appear to be connected. Rather, he saw them as collections of events meaningfully related in significance, but not causally connected. He wrote a book about synchronistic events in which he claimed that life is not a happenstance of random events, but rather directed manifestations of an innate order of psychic phenomena connected to the collective unconscious. In other words, his synchronicity is the simultaneity of time with space and mind where something other than chance is involved. As an example, Jung tells us that someone might notice that the number on his theater ticket is the same number as one on a bus ticket he bought that same day. The point is that the coincidence is in the noticing that the two numbers are the same.

The person "chanced" first to notice and remember the number, which is already an unusual thing to do. What caused him to notice the number? Jung tells us that there may have been some kind of "foreknowledge of the coming series of events."[5] He tells us that cases like these, in every conceivable form, happen frequently, but after the first momentary astonishment they are often quickly forgotten. Jung would say that there is some sort of elevated archetypal phenomenon going on at the time a person notices a critical event. Extraordinary connections link one more closely with the archetypal universe and therefore bring more of a participating connection between the subconscious and the conscious. I agree with Jung in believing that the marvels of coincidence are in the connections between participatory foreknowledge and awareness.

There is a wonderful exchange of letters between Jung and Wolfgang Pauli (on Jung's theory of "causeless order").[6]

Pauli was a physicist. And for physicists, events usually have causes. I say usually, because the physics of relativity and quantum theory have those bizarre associations that appear to have no cause whatsoever. That's because those atomically small particles don't behave like the larger ones governed by natural laws of cause and effect. The behavior of those very small particles (if we can call what they do "behavior") is known only through statistical truths and predictions, not through ironclad linkages between cause and effect. In Jung's example, of someone buying a theater ticket that bears the same number as a train ticket he bought on his way to the theater, we have a clear pairing of two events that are not likely to have a knowable cause. Indeed, our days are full of such pairings. We just don't notice them. Every now and then we become more alert to such pairings. Jung gives the example of the pairing of the word and notion of *fish*.

> I noted the following on April 1, 1949: Today is Friday. We have fish for lunch. Somebody happens to mention the custom of making an "April fish" [April fools] of someone. That same morning I made a note of an inscription which read: "Est homo totus medius piscis ab imo." In the afternoon a former patient of mine, whom I had not seen in months, showed me some extremely impressive pictures of fish which she had painted in the meantime. In the evening I was shown a piece of embroidery with fish-like sea-monsters in it. On the morning of April 2 another patient, whom I had not seen for many years, told me a dream in which she stood on the shore of a lake and saw a large fish that swam straight towards her and landed at her feet. I was at this time engaged on a study of the fish symbol in history. Only one of the persons mentioned here knew anything about it.[7]

Jung claimed that the sequence of fish events made quite an impression on him, mostly because it was exceedingly odd for all those fish happenings to recur on the same day. It was what he meant by a *meaningful* coincidence, something he seemed to define as an *acausal* connection that is very natural. Of course, we should remember that in Jung's day, it was quite usual for many people around the world, especially Catholics who were not permitted to eat warm-blooded animals on Fridays—presumably because Jesus died on a Friday—to associate Fridays with fish. So, there is one causal relationship. And on April 1, at a time when the name of April Fool's day was April Fish, Jung would have been thinking of fish. Besides, Jung admitted to having been working on the archetypical symbol for fish for months before his April 1 fish event. That would also contribute to the cause of being aware of any notion of fish whenever it came up since they are naturally archetypical symbols. So, Jung's fish connections might just be causal. On the other hand, they might be related simply by what Jung calls *meaningful cross-connections*.

Jung set out to build a theory of the mind that parallels that of space-time, a theory that has no need for causal order, a theory where chance seems to drive the connection between two events. Just as Einstein added time to space to produce the much deeper concept of relativity, so Jung proposed completing causality by adding a noncausal link.[8] Certain patterns, he argued, are linked in nonmechanical ways to form a "causeless order." . . . its patterns are meaningful and are echoed in both mind and matter.[9] To Jung it was a psychic energy, as if there is some energy field coming from the collective subconscious of significant experiences within the mind—not the neural electrochemical energy whirling about the mind, but rather a kind of energy current of archetypes of the unconscious that connect significant experiences. Could there be such energy, an energy of meaning without a cause, an energy of synchronic psychic events exciting some archetypal connection?[10]

Jung's position on meaningful coincidences is persuasive. He held that meaningful coincidences create powerful undercurrents in a person's psyche and that consequential synchronic events of the conscious being are interconnected with the subconscious.

Coincidences tie us to the intricacies of life, uncover a sense of ourselves, and give meaning to our own existence. A coincidence, such as a double rainbow believed to be a sign from the dead, gives meaning to the notion that we are all forever attached to the people we care about. Its archetypal connection is the rainbow itself as the symbol of a door open to heaven. We see connections to the larger universe the moment we encounter a coincidence. Even a simple connection makes us feel a part of the galaxy, and possibly beyond. Most days we go about our lives without noticing any connections, as if there were no web out there. We hardly realize that there are more connections just around the corner from where we are. We seldom recognize synchronic connections staring us in the face and are surprised when we see them, and therein the delight.[11] But reactions to surprise in true stories depend on how they are told. Specific details can make the same coincidence more surprising and more meaningful when told as a predictor of future events instead of as if it had just happened. A personal story is bound to be more surprising and more meaningful to the person telling the story than to the person listening to the story. To me, the story of the albino taxi driver did not seem as surprising, and certainly not as meaningful, as my bumping into my own brother after hearing his familiar laugh in a café on the Bay of Mirabello in Crete. The stories in the previous chapter are striking, yet also inevitable in long lifetimes.

Over the last few years I've heard many coincidence stories that on first thought are astounding. Some involve mistaken identities. Some are about being in the right place at the right (or wrong) time. They include, but are not limited to, chance meetings and physical accidents. Others are about winning (or losing) at games that depend on random events. And still others involve telepathy and clairvoyance. Most can be explained, more or less, by simple mathematics accounting for a higher likelihood than one expects. The stories are astounding only when viewed through statistical misconceptions or by underestimating (or overestimating) the size of the world and its population. Why is it that we all have so many stories that can be classified as similar to many of those in the previous chapter? The answer can be simply explained with a little knowledge of probability, and of how it works in nonintuitive ways.

PART 2

The Mathematics

Collisions

In a world so large or small
there are events we never see
that sooner or later happen.
If in neither a thousand nights
by some odd circumstances,
nor twice in a million months
of the moon's waning crescents,
nor ten thousand leap years,
of surprising perchances,
then I assure you they will
after earthshaking dances
of large numbers with chances.

—J. M.

HERE WE PRESENT a few mathematical tools for examining coincidence stories—the law of large numbers, the law of truly large numbers, the birthday problem, some probability theory, and some frequency distribution theory. This part covers all the mathematics that are handy to know to be able to understand the central argument of the book, which is essentially: If there is any likelihood that something could happen, no matter how small, it's bound to happen sometime. Some of the mathematics will be used to analyze the stories presented in Part 1 and returned to in Part 3.

Chapter 4

What Are the Chances?

> What I found were "coincidences" which were connected so meaningfully that their "chance" concurrence would represent a degree of improbability that would have to be expressed by an astronomical figure.
>
> —CARL GUSTAV JUNG[1]

WILDLY ASTONISHING COINCIDENCE stories invariably conclude with the question "What are the chances of that?" It's generally meant to be rhetorical because it's never an easy question to literally answer. Although there are basic statistical techniques and good-practice experimental models for studying the rarity of coincidences, mathematicians do not have an adequate broad theory of the subject as yet. The problem lies in the definition of the word itself. After all, *coincidence* implies an accident with no apparent cause, which includes flukes and miracles. And what would we do without the hope and glory of miracles? Perhaps measuring the odds of coincidence is an oxymoron. How can we know the odds of a happening that has no apparent cause? One might argue that the roll of a pair of dice landing on boxcars has no apparent cause, other than a hundred inestimable variables that determine their flight, and yet we are able to give the outcome odds of 35 to 1 against. We have an actuarial grip on the odds of a person living past the age of x years. So, what is obstructing us from measuring the odds of a miracle or that a dream of meeting a tall dark stranger across a crowded room will come true? We don't always need to know a cause of an event to get a handle on a measurement of its odds. We didn't know why cigarette smoking caused cancer before we learned that it did by processing the statistical odds of someone getting the disease. That happened after WWII when women, who had not been smokers before the war, joined the workforce and started smoking. Up went their cancer rate, and bingo: We surmised a correlation and connected the dots. The problem with many coincidences lies with the enormous number of variables that cannot really be known or inferred from statistical samples. Coincidences are not easily explained by quantitative analysis; yet there are qualitative reasons to suggest that they happen more frequently than we expect. Even psychics avoid quantitative predictions in preference of the qualitative.

When we think about coincidences we think about likelihood. Tell a coincidence story and someone inevitably asks, "What are the chances of that?" The answer almost always engages words that mean "pretty slim." It is the job of people in probability theory to tell us the meaning of *pretty slim*, or at least, to think about what it means. A measure of an event's likeliness is a number that mathematicians call a *probability*. It is always a number between 0 and 1, where 0 indicates impossibility and 1 indicates absolutely certainty. There are several ways to get that measurement. One is to look at relative frequencies from a large sample. In principle, the probability of an event is a ratio of two numbers, each of which can be determined by observing the proportion of repetitions by which the event occurs. As the number of trials increase, the relative frequency of an event approaches the probability of the event. A second way to measure probability is to count the logical possibilities: a rolled fair die can land on any one of just six sides. We don't need to roll the die to know that the probability of rolling an even number is 1/2, or 50 percent chance.

If two events are connected in such a way that they cannot both happen together because of some logical constriction—such as drawing a red queen *and* a queen of spades on a single draw from a standard deck of fifty-two playing cards—then the probability that one *or* the other will occur is the sum of the probabilities of each event; in other words the probability of drawing a red queen or a queen of spades is $1/26 + 1/52 = 3/52$.

The general idea works like this: Suppose X represents the outcome of an event, and $P(X)$ the probability that the event actually happens. Then the probability that the event does not happen is $1 - P(X)$. For example, in case you were flipping a coin, $P(heads)$ would equal 1/2, and so would $P(tails)$. Or in case you were flipping a pair of dice, $P(4) = 1/12$ and $P(not\ 4) = 11/12$. If X and Y are possible outcomes that are independent from each other (that is, one has no effect on the chances of the other), the probability of X *and* Y happening is the product $P(X)P(Y)$, and the probability of X *or* Y happening is the sum $P(X) + P(Y)$.

For an example of a human coincidence, take one event to be accidentally bumping into your best friend in Bora Bora next Tuesday morning and another to be accidentally bumping into a cousin in Reykjavik that same Tuesday afternoon. The first event has an effect on the second. Unless you have access to an F-15, you cannot accidentally bump into your best friend in Bora Bora in the morning *and* accidentally bump into a cousin in Reykjavik on the same afternoon. Naturally, allowing two possibilities gives a better chance. In the case of cards—a red queen could be drawn, *or* a (black) queen of spades could be drawn. If, on the other hand, we have a situation where one event is completely independent of another, then the probability that both will occur is the product of the probabilities of each event. The probability of drawing a red queen once, and then a queen of spades after replacing the red queen would be 1/26 + 1/52 = 1/1352.

Indeed, demanding that two prescribed events must happen gives less of a chance. On the other hand, the probability of drawing both from a deck without replacing the first drawn card makes things a bit more complicated. We would be asking for the probability that one event will happen after another event already happened: a *conditional probability*. The case of drawing two cards from a single deck is instructive. If we assume that the drawn card is not replaced, then the probability of drawing a red queen and then the queen of spades would be 1/26 + 1/51 = 1/1326. At the time of the second draw, the deck is missing one red queen and is therefore down one card. So, the probability of drawing a queen of spades on the second draw would simply be the probability of drawing it from a deck of fifty-one cards. By not replacing the red queen, the likelihood of drawing the queen of clubs is increased. What is important here is that we are dealing with a product of two numbers less than 1, which means that the resulting probability will be less than the probability of either event. To complicate things a little, notice that we stipulated that the queen of spades was drawn after the red queen. Had we asked for the probability that either card be drawn—spade first or second—the probability should be greater. We would then consider two probabilities: the probability of drawing the queen of spades followed by a red queen and the probability of drawing a red queen followed by the queen of spades.

The Difference Between Odds and Probability

We distinguish between the terms *odds* and *probability*. When we say the odds are m to n, we mean that we expect the event to not occur m times for every n times it does occur. The standard notation $m{:}n$, which in words translates as *m to n*. If the odds are m to n, the probability is simply the ratio $n/m+n$, so 4-to-1 odds converted to a probability is 1/5. To compute the odds of an event that has a probability of success p, compute the ratio $(1-p)/p$, and reduce it to m/n. Then the odds against the event happening is *m to n*. In the case $p = 1/5$, the ratio becomes $(1 - (1/5))/(1/5) = 4/1$, and so the odds are 4 to 1.

The idea of odds comes from gambling. It's easier to compute winning; a winning bet of $1 paying m to 1 would collect $$m$, an amount that already includes the original stake. Even odds or even money means the odds are 1 to 1. In this book we will try to reduce odds to the case where $m = 1$. Conceiving likeliness or unlikeliness is easier when we know that there are m failures for every 1 success. On occasion we will use the expression "chances are 1 in m" to mean that there is one way to succeed in m trials. So, for example, "the chances of drawing an ace of spades from a deck of fifty-two cards is 1 in 52" translates into "the odds of drawing an ace of spades from a deck of fifty-two cards is 51 to 1."

A Probability Thought Experiment

Pick any two events that might have some slim chance of happening. Take the first to be that a black cat will walk across your path next Wednesday. For the second, suppose that sometime in your life you will get a registered letter from a law firm, saying that great-uncle you never heard of passed away and left you a million dollars. Suppose the first event has a probability of 0.000001, given the population of black cats roaming the streets in your neighborhood. Suppose that the second has a probability of 0.000001, given that you don't have too many uncles that are not known to you. (I'm making up these numbers for the sake of the argument.) The probability that both will occur is extraordinarily small, just 0.00000000001. It is a smaller probability than that of either event's happening by itself, and a larger probability than that of both happening at the same time. Of course, the probability that one *or* the other will occur is greater.

Now consider ten distinct rare events—

a. A black cat walks across your path on a Wednesday.
b. A great-uncle you never heard of passes away and leaves you a million dollars.
c. A ring you lost twenty years ago turns up in a garage sale on your street.
d. A dream of meeting a tall, dark stranger across a crowded room comes true.
e. You play the Texas Lotto lottery, and win twice.
f. You meet your own brother in Bora Bora by coincidence.
g. In a foreign bookstore you find a copy of Mark Twain's *The Mysterious Stranger* with your name penned on the title page.
h. You renew your passport and the new passport number turns out to be your social security number.
i. On a park bench you find a copy of Mark Twain's *The Mysterious Stranger* that belonged to you when you were a teenager (yes, an event very similar to *g*).
j. You call for a taxi in Chicago and recognize that the driver is the same person who drove a taxi you called for in New York a year ago.

I picked these arbitrarily. Some are coincidences and some are just singular events. They could be completely independent events, were it not for that meddlesome old proverbial butterfly over the Pacific—the one that seems always to have an effect on everything from the weather in Paris to the results of the Kentucky Derby—that seems to always cause some unexpected trouble. Why did the black cat appear at that particular moment? The tall, dark stranger might be just the guy who finds your long-lost ring brought to him by the black cat.

Probabilities for some of these events and others of their kind are extremely hard to know—even approximately. For the sake of simplicity, suppose each of these events has a probability 0.000001, a number less than the probability of being dealt a royal flush on a single deal. There is no special reason for picking that number, other than the fact that it tells us that the event is not impossible and not likely. It might seem as if the probability that one of two of the events on the list will happen is 2 × 0.000001 = 0.000002, because probabilities are added when calculating the probability that one of two events will happen. That would naively suggest that the odds double by considering just two possible events. But we must be careful. The calculation ignores the possibility that both events (such as *g* and *i* on the list) might depend on each other. So, we must subtract the probability that both will occur: 0.000001 × 0.000001 = 0.00000000001, a relatively small number. The actual probability would then be 0.00000199999, slightly smaller than double. This leads us to a curious question. The answer might have us view the coincidence world differently than before. In the world of all the possible fantastically surprising happenings, there must be thousands—perhaps millions, or billions—that might happen to you in the course of one year. Let us assume the probability of each of a million such happenings be, say, 0.000001. Now, the question is this: What happens if we group all those happenings and ask for the probability that at least one will happen within a year? There is no practical way of determining the independence of a million happenings. We cannot assume that no two happenings have any direct connection. We cannot discount the possibility that one event might cause or influence another, or that one singular event might depend on another. For example, if you won the lottery once, that might influence you to spend some winnings on trying again, so winning a second time then depends on your first win. And so we cannot simply add the probabilities together to get the probability that one of the 1 million happenings will happen. That would lead to the absurd calculation that the probability of one happening is 1,000,000 × 0.000001 = 1, or certainty! (We would be adding 0.000001 to itself a million times.) To make the calculations work, events should be disjointed, have nothing in common. If they do, any serious measuring of probabilities gets exhaustively complicated, or impossible. For instance, we would have to eliminate the possibility of the black cat that might walk across your path next Wednesday will find your long lost ring in a drainpipe and bring it to the tall, dark stranger, who will sell it in a crowded tag sale. But even then, even when all these requirements are met, we will still have to account for an enormous number of intersecting possibilities that would shrink the odds. On the other hand, if those million happenings were independent, then the mathematics would tell us that we could be certain that one of them will happen. Of course! Any active person would encounter one of a million things that could happen. By just leaving the house, a person encounters an enormous number of possibilities.

Event *e* is the only one on our list that has a fairly precise probability, though even that event depends on the personality of the winner. To win twice, you have to first win once. That means first picking the right six numbers. The probability of that happening once is close to 0.000000038, a very small number indeed.[2] Another way of putting it is that your odds of winning would be 25,827,164 to 1.

How is this calculated? There are 54 possibilities to pick one number. Once the first number is picked, it is not replaced, so there are just 53 possibilities in picking the second number. Likewise, there are 52 possibilities for the third, 51 for the fourth, and 50 for the fifth, and 49 for the sixth. So, there are 54 × 53 × 52 × 51 × 50 × 49 × = 18, 595,558,800 different ways of choosing six numbers, each from 1 to 54. There are 1 × 2 × 3 × 4 × 5 × 6 = 720 different orderings of six numbers. Since the order in which the six numbers are picked does not matter, we divide by 720 to get 25,827,165, the number of different possible pickings, only one of which is correct.

The probability of winning the second time remains the same; the lottery numbers have no memory and neither does the probability. That probability, however, depends on how we think about it. If you forget the fact that you won the first time, then the probability does not change. Your odds are still 25,827,164 to 1 with a probability of 0.000000038. The probability of winning twice is $0.000000038 \times 0.000000038 = 0.000000000000001444$, suggesting that winning twice is very, very unlikely. We know that the winning lottery number has no history. However, in a strange way, the winning itself has a history based on the personality of the winner. Like criminals returning to the scene of their crimes, winners return to playing the lottery. They continue to play with deep pockets, buying far more lottery tickets than they ever had before. So, our calculation ignores all other attempts at playing the lottery. A person could play a hundred times before winning the second time. In Chapter 7 (specifically, in Table 7.1), we will find odds of winning the lottery four times in exactly four tries, a far harder thing to do.

Chapter 5

Bernoulli's Gift

How is it possible to have a mathematical law that tells us something about the future? After a pair of dice is rolled once and picked up, the dice "forget" where they landed before. If the dice are fair and they are rolled without cheating, we cannot tell in advance the outcome, and yet we can be pretty sure that, over many rolls, 7 will appear more often than any other number. It's a matter of the geometry of dice and a simple assurance of arithmetic: more pairs of the numbers 1 through 6 add up to 7 than to any other number on which a pair of dice could possibly land.

The mathematics of probability is relatively new. It doesn't go back much further than the sixteenth century. Before the beginning of the sixteenth century, mathematics did not deal with uncertainty. Natural philosophers and mathematicians were more interested in understanding the serious things in life, which to some were the abstract notions of number theory and geometry and to others were the more practical and functional things in life, such as surveying and other building practices (especially for cathedrals). The whole mathematical idea of chance should have emerged from Girolamo Cardano's *Liber de Ludo Aleae* (The Book on Games of Chance), a folio of papers containing the essential elements for understanding the nature of chance and modern probability, which was written close to the year 1563.[1] But *Liber de Ludo Aleae* remained unpublished for the next hundred years.

Girolamo Cardano was a Milanese physician, mathematician, and a gambler. We know him mostly from his book *Ars Magna* (The Great Art) published in 1545. It is an account of everything known about the theory of algebraic equations up to that time. The *Liber de Ludo Aleae* was fifteen pages of rambling mathematical and philosophical jottings. Cardano had no intention of publishing it. But in the *Ludo Aleae* we find some handy tools for studying frequencies of coincidences. We now consider it to be a cornerstone of probability theory, expected value, averages and means, frequency tables, the additive properties of probabilities, and calculations on the combinations of ways of having k successes in N trials. It even contained a tip suggesting a mathematical law that would later become known as *the weak law of large numbers*. Roughly, the law tells us that the difference between the actual observed probability (which is entirely unknown before the events happen) and the mathematically calculated mean p is likely to be as small as one wishes, provided that the number of trials N is large enough.

When expressed in its precise form, it is an enigmatic mouthful: the probability P that the average success ratio differs from p is as close to zero as one wishes, provided that N can be taken as large as needed to force that condition. In modern notation, where ε represents any small number chosen, $P\left[\left|\frac{k}{N}-p\right|<\varepsilon\right]$ converges to 1 as N grows large.[2] For those readers who might have jumped when seeing this last cocktail of symbols, let me explain. We are using notation designed to talk about the probability of an event described within the square brackets. For example, P[*A hurricane hits Central Park next July 4*] denotes the probability that a hurricane will hit Central Park next July 4. So, $P\left[\left|\frac{k}{N}-p\right|<\varepsilon\right]$ just denotes the probability that the absolute value of the difference between the ratio k/N and p will be less than any small number ε chosen.

It is a principle that suggests how averages are likely to behave in the long run. One must wonder how it is possible for random events (with absolutely no history of each outcome) to have a mean close to a mathematically calculated number. Unfortunately, this magnificent and true law—even today—is often confused with what some people call the *law of averages*, which isn't a law at all, but rather a nonsensical phantasy that says if you flip a coin enough times, half the time it will come up heads and half the time it will come up tails. Unless we take "long enough" to mean infinitely long, this "law" is not at all true.

Yes, the weak law of large numbers is a truly astonishing result. But even more astonishing is that it can be proven mathematically! It demonstrates that it is possible for random events—events with a possibly wide variation of possible outcomes and with absolutely no history of each outcome—to have an observed average close to a mathematically calculated number. Mathematics can tell us about determinant phenomena of the real world—the structures of bridges and dams that obey mathematical computations. Planes fly and windows break according to mathematics. Glass breaks at certain resonant frequencies; plane airfoils lift when the pressure above is lower than the pressure below. But when it comes to chance, the connections seem far more mystifying. Dice? How could we possibly know which way they will fall on any given throw?

Cardano posthumously gave us a way. Before his *Liber de Ludo Aleae*, luck—good or bad—was in the hands of Tyche, Fortuna, or some other divinity that caused chance to favor one outcome over another. Even the Greeks, who excelled in so many areas of remarkable mathematics, had no mathematical theory of gambling odds. They simply cast their dice, believing that either luck, fortune, or some god determined their fates. Oh, sure, they knew that certain numbers were more likely than others to turn up. Undoubtedly, they knew that 7 would come up more often than any other number. All they had to do was count the number of ways a 7 could come up against any of the ways other numbers could. But, as far as we can tell, they had no notion of predictive odds.

Cardano's little manuscript held the seeds and secrets of the science of chance. We learned that observable facts could quantify what is likely to happen. According to Henri Poincaré, the world then learned that a person has the same chance as any other person and even the same chance as the gods.

We must remember that in Cardano's day, there was no well-studied notion of simple explanations of chance. For instance, mathematicians were not thinking of reasons for why some numbers appear more often than others. Galileo solved that mystery half a century after Cardano's death when he wrote a short treatise on the odds of throwing three dice, although it's not likely that Galileo knew of Cardano's *Liber de Ludo Aleae*. He listed all the combinations and found that there are twenty-seven distinct ways for three dice to add up to 10 as well as for them to add up to 11, but only twenty-five ways for three dice to add up to 9 or 12.[3]

Surely, experienced gamblers already knew this. They had a fundamental understanding of dice outcomes from their folk knowledge coming from centuries of practice and observation. They had an instinctual knowledge of the odds, and knew that for three dice, 10 and 11 turn up more frequently than any other number. But there is a difference between having those instincts and having mathematical explanations. With the confidence of mathematics, you could almost count on your luck. For those who knew how to calculate mathematical odds, decisions were no longer such a risk. In the long run, they were *almost* a certainty, notwithstanding those random pinches of uncertainty brought forth by flukes and coincidences.

Double Sixes and the Birth of Probability

The core ideas of mathematical probability can be traced back to the winter of 1654. It was an unusually cold winter in Paris. Even the Seine froze. It was reported that Parisians skated on the river while fires burned at street corners where parish priests distributed bread to the poor. The economy was stifled by thirty years of European religious wars that had drained the French treasury. France was forced to increase its taxes on the working class, but dishonest tax collectors brought little revenue into the treasury. Louis XIV was king and the nobility, being exempt from taxation, was accumulating appalling excesses of wealth. It was no coincidence that the idle rich were overtly gambling in gaming rooms all over Paris.[4] And it was no coincidence that a burgeoning mathematical theory of probability appeared at that time, indeed in the winter of 1654.

Although gambling goes back to the beginning of time, or at least as far back as when cavemen rolled bones, by the mid-seventeenth century it had become the centerpiece of pastime entertainment in France. There was still no serious mathematics of chance, apart from the few crude attempts found in some erroneous arithmetic texts and the Franciscan Fra Luca Pacioli's *Summa*, published in 1494, a textbook mostly on algebra. But by 1654 the manuscript of Cardano's *Liber de Ludo Aleae* had surfaced with some clues to the smallest number of times a person must throw a pair of dice to have a better than even chance of getting a double six.[5]

Mathematician-philosopher Blaise Pascal read a copy of the *Liber de Ludo Aleae* in search of that number, but didn't believe its solution. He became ill, and confined to his bed during the spring and summer, corresponded with his friend, lawyer-mathematician Pierre Fermat.[6] Together, they concluded that the odds are slightly less than even that double six would turn up on twenty-four throws and slightly more than even on twenty-five throws.[7]

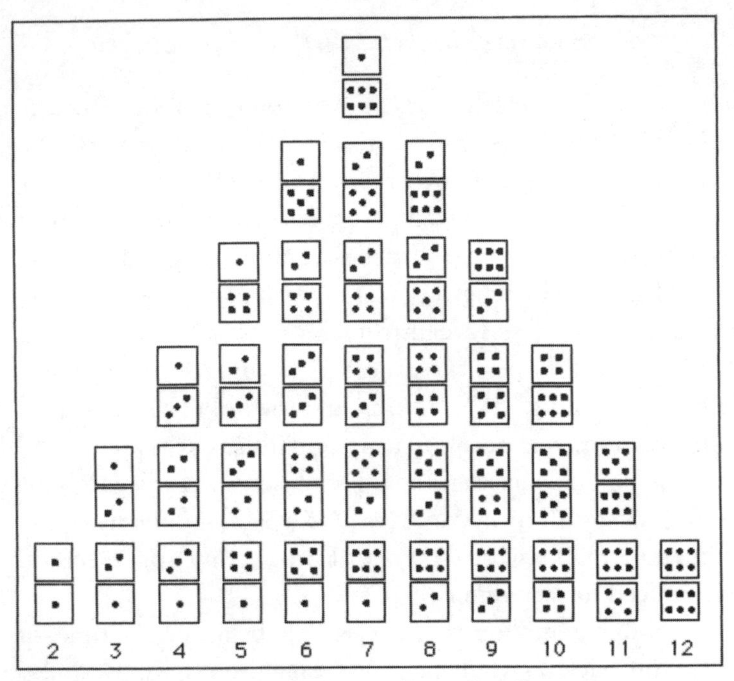

Figure 5.1. The number of pairs in each column represents the number of ways each number can appear.

Pascal knew that snake eyes (double ones) and boxcars (double sixes) very rarely turn up since they have a 1-in-36 chance of doing so, whereas seven has a 1-in-6 chance (see Figure 5.1). He understood that it would be easier to calculate the likelihood of *not* throwing a double six, which turns out to be $1-1/36$, or 35/36. He also understood that each throw is independent of the previous throw, and that the probability of two independent events is the product of the probabilities of each event, and so, the probability of *not* throwing a double six on n throws is $(35/36)^n$. He calculated $(35/36)^{24}$ to be 0.509 and $(35/36)^{25}$ to be 0.494, to conclude that there is a slightly smaller than even

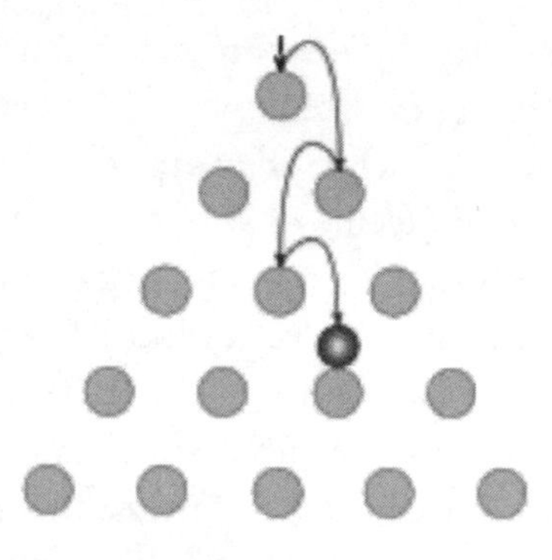

Figure 5.2. Galton board: looking at fifteen rods perpendicular to the page.

chance of getting double sixes on twenty-four rolls of the dice, and a slightly better than even chance with twenty-five rolls.[8]

The foundations of probability came from this dice problem and other similar problems. The outer layer of so much of the stochastic world can be encapsulated by one illustration. Think about the world this way: If an event is influenced by some cause, there is a better than even chance that the cause will lend favor to a direction of the event's future. If an event is not influenced by any cause, the direction of the event's future could go one way or another without bias. Cause or not, a better than even chance leaves a door open to the unpredictable winds of fluke or coincidence. In Figure 5.2 we illustrate this by using a so-called Galton board as model.

The Galton board models behavior that is decided by impartial chance. A ball is dropped on a set of rods in such a way that it hits the first rod so precisely at the top of the rod that it has an exactly even chance to bounce to the left or to the right. If it favors the right side, then it will descend to the next lower rod and, once again, will have to either hit the top of the rod, or favor one side or the other. In theory, such a ball can hit the precise top of a rod. In practice, however, it never happens. Why? First, we must consider what the top of the rod really means. Does it mean the top molecule of steel (assuming the rod is made of steel)? There is no such thing. So, in practice there are causes for the ball to favor one side or another. Perhaps those causes include the minuscule current of air the ball must go through, or the minuscule vibrations coming through the mountings of the rods, or the smallest piece of dust particle that got hit in impact. In practice there are hundreds of variables that determine which way the ball will bounce after its impact with the rod. In addition, a molecular dent and the elasticity of the collision have to be taken into consideration.

Nineteenth-century English geneticist Sir Francis Galton constructed such a board of pegs arranged quincunically, like the dots on a die's 5 face. Galton's point was to demonstrate that physical events ride on the tailwinds of chance. In the absolutely perfect Galton board, as one in which the balls always fall precisely on the absolute tops of pegs, the ball falls to the right or left as if it flipped a coin to decide. In real life, a butterfly flapping its wings over the Pacific or a cow farting in an Idaho cornfield might determine that decision. Before each bounce, the outcome of the previous bounce is forgotten history; the ball no longer remembers the outcome and therefore behaves as if it had just hit the first peg. And yet the cumulative outcome seems to take into account the history of all previous outcomes.

Let us look at this mathematically. Suppose the ball hits four layers of pegs on its descent. The even chance of going left or right causes the buildup of balls below the rods in the shape of a bell-shaped curve. Counting the number of ways the balls can fall proves this. Suppose that a ball is dropped and we mark its descent by the letters L and R to indicate bouncing to the left or right. We would then have the following possible outcomes:

LLLL
LLLR, LLRL, LRLL, RLLL
LLRR, LRLR, LRRL, RLLR, LRLR, RRLL
LRRR, RLRR, RRLR, RRRL
RRRR

There are more combinations of mixed letters than non-mixed, and since there is an equal chance for the ball to go left or right, there is a tendency for it to favor the center region, under the top rod. The reason for this is that, in a string of, say, twelve choices of L and R (as in Figure 5.3), there are more strings with six L's and six R's than any other number of L's and R's.

At each impact with a rod, count the falling of a ball to the left as –1 and its falling to the right as +1. After bouncing down twelve rows of pegs, the ball will end up in one of the twelve pockets at the bottom of the board.

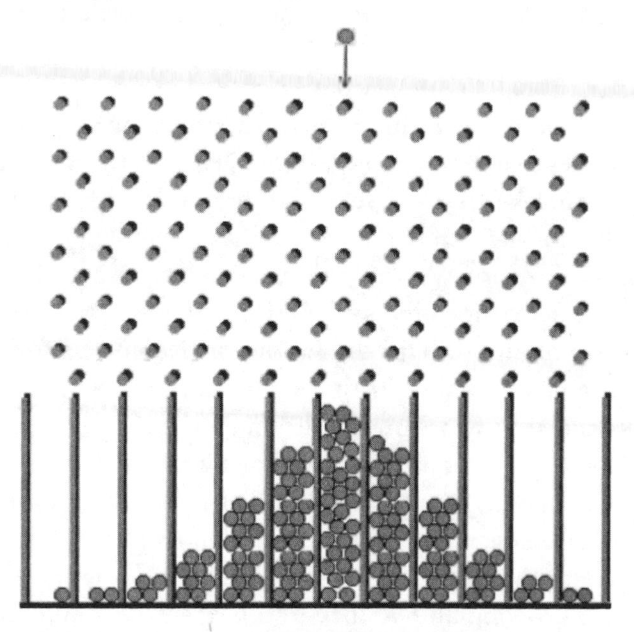

Figure 5.3. One hundred forty balls falling through a Galton board.

So, for example, the ball at the extreme left in Figure 5.3 will end up with a cumulative value of –12. The final position of each ball represents a distinct cumulative outcome. The balls tend to accumulate toward the center. However, though quite a few balls fall in the two center slots, more fall in the ten remaining slots.

In Figure 5.3, the collection of balls represents the final accumulated values of 140 experiments—31 fell in the five slots on the left; 55 into the five slots on the right; and 54 into the two middle slots. It is true that the final position of any one ball does not indicate the history of its journey. However, notice two critical things: (1) the first two rows of pegs limit the outcome; and (2) heads on the first and tails on the second (or vice versa) force the final accumulated value to be less than 12 and greater than –12. Almost 60 percent of the balls have fallen outside the center two slots. Now, it is possible for a ball on the left that has fallen a few layers to end up on the right, but it is also true that any ball that wanders too far to the left will have a decreasing chance of returning to the right.

Today, probability theory develops in empirical and abstract directions. For instance, the empirical approach would be to use large samples to estimate a probability, whereas the abstract approach would be to use a scientific principle to lock down a probability through known facts, such as a symmetry argument or a physical theory. We know the probability of a perfect die's landing on 1 by the cubic symmetry of the die itself. But the probability of an ordinary die falling on 1 could be found by rolling it a large number of times and marking the number of times it landed on 1; its probability might turn out to be more or less than 1/6—after all, it is a real die with real imperfections.

Much depends on the die itself. The dice that come with packaged board games are crudely manufactured. Yahtzee is a dice game that has been around since the 1950s. It is played with five dice. Throwing all five dice and having them all come up with the same number is called a Yahtzee. The odds of throwing a Yahtzee are 1,295 to 1.[9] You might expect that it would take 1,296 tries to throw that Yahtzee. But if many people all over the world spend some time trying, it might easily happen on the first try. That's exactly what Brady Haran thought when he asked his hundreds of followers on his website to try for a Yahtzee and video record their Yahtzee shot. And wouldn't you know that some people threw Yahtzees after just a few throws, and many were successful after just a few hundred throws.[10]

In the eighteenth century, to find the likelihood of an event, you would simply count cases: you would take the ratio of the number of hoped for outcomes to the number of all possible cases. A fair die could fall on one of six possible faces, so the probability p of that die falling on any particular face is 1/6. But Bernoulli asked the question differently. He wished to extend it to include problems involving disease and weather, with the hope to include other scientific questions.[11]

Bernoulli's Theorem

Mathematicians are frequently awestruck by the magnificence and beauty of an abstract principle. They are moved by the kind of beauty that emerges when theory is elegantly applied to the natural world. Swiss mathematician Jacob Bernoulli was jubilant when he proved the weak law of large numbers after seeing Cardano's *Liber de Ludo Aleae*. That law is truly amazing because it tells us that though nature is unpredictable, with its unfathomable number of ingredients and variables, we still have fantastically clever ways to measure its secrets.[12] It gives us an amazing handle on uncertainty.

When Jacob Bernoulli died in 1705, he left reams of incomplete and unpublished manuscripts to his nephew Nicholas Bernoulli. For the next eight years Nicholas worked through his uncle's papers and finally published *Ars Conjectandi* (The Art of Conjecture), a groundbreaking work that even today is recognized as presenting some of the most critical early notions of the mathematical theory of probability. Posthumously published in 1713, the *Ars Conjectandi* took a unique approach by giving the example of an urn filled with white and black tokens and telling us how to tell the ratio of white to black, even when we do not know that it contains three thousand white tokens and two thousand black. First, understand that there is a mathematical probability given as the ratio of the number of white tokens to the number of all tokens. But we do not know what those numbers are. So, how can we know that mathematical probability? Here's Bernoulli's plan: You blindly choose one token, record its color, put it back, and shake the urn. If you repeat this, blind picking tokens one by one for a sufficiently large number of times, you will get pretty close to that secret mathematical probability. In fact, even as you increase the number of picks, you will come closer to that mathematical probability. Suppose, for instance, that after 200 blind picks, you have recorded 120 whites and 80 blacks. Then, the number of white to black will be in the ratio of 3 to 2. You could then assume that the probability of picking a white is 120/200, or 3/5.

Bernoulli's *Ars Conjectandi* gave us the weak law of large numbers. For tossing a fair coin N times with the hope to have k heads, the theorem tells us something about the probability of how close the ratio k/N will be to 1/2, the mathematical probability that heads *will* turn up on a single toss. By some confused wishful thinking, many gamblers take it to mean that for high values of N the outcomes of events will come close to the probabilities of those outcomes. Thus, to take coin flipping as an example once again, the confusion suggests that since $p = 1/2$, the total number of heads will converge to the total number of tails over the long run. The theorem says only that the *likelihood* of that happening is converging on certainty over the long run. There are no guarantees of what would happen in any individual case. As an example, let us suppose that we have a game of N repeated events, such as flipping a coin N times, and we count the number of times the coin lands on heads. The mathematical probability of a fair coin landing on heads is 1/2. What will we actually observe when we flip a coin in real life? Will the success ratio k/N be close to 1/2; so close, say, to be within a difference of 1/10,000? We cannot really answer that, but we can put it another way and ask whether there will ever be a time when the probability will be greater than, say, 0.999 that the difference between k/N and 1/2 is less than 1/10,000. Bernoulli's theorem says *yes, there will be such a time*, if N continues to increase with time. But it does not totally prohibit occurrences when the difference between k/N and 1/2 is greater than 1/10,000 even for large N. In fact, even if the success ratio k/N comes close to 1/2, there is no assurance that it will continue to be close. Moreover, it turns out that a slightly stronger version of Bernoulli's theorem tells us that although the success ratio k/N is likely to converge on 1/2, the actual success values tend to behave increasingly wildly. Consider this surprising statement: the probability that the actual number of successes deviates from the expected number $k/2$ of successes (that is, landing on heads) becomes more and more likely as the number of trials grows very large. Although counterintuitive, it is true.[13] However, it also says that in the long run the difference between the actual mean that we can get empirically after the trials (which of course is entirely unknown before the trials occur) and the mathematically calculated mean is likely to be as small as we wish, provided that the number of trials N is large enough. It means that random empirical events (with absolutely no memory of each outcome) have a mean close to a mathematically calculated number!

Bernoulli was so happy with his theorem that he imagined it applying to the most general events for everything in the world. In his *Ars Conjectandi*, he wrote:

> Whence at last this remarkable result is seen to follow, that if the observation of all events were continued for all eternity (with the probability finally transformed into perfect certainty) then everything in the world would be observed to happen in fixed ratios and with a constant law of alternation. Thus in even the most accidental and fortuitous we would be bound to acknowledge a certain quasi-necessity and, so to speak, fatality. I do not know whether or not Plato already wished to assert this result in the dogma of the universal return of things to their former positions [*apocatastasis*], in which he predicted that after the unrolling of innumerable centuries everything would return to its original state.[14]

In theory, Bernoulli's theorem should have been an intellectual explosive, a tour-de-force mathematical measure of uncertainty. It promised to predict the future. It is where we first encounter a mathematical law that gave immense yet straightforward insight into how chance behaves in the real world, a theorem that Bernoulli proudly announced as hard, original, and so excellent that it gave dignity to all parts of his treatise. But Bernoulli was discouraged by some of his own experiments that had applications to problems involving disease and weather. He ambitiously gave himself an extremely harsh criterion of certainty, even by today's accepted standard.[15]

Bernoulli has given us enormous powers applied to the uncertain behavior of nature as well as to games of chance, a method of finding the expected value from no a priori information. "And indeed, if in place of the urn we substitute, for example, the air or a human body, which contain within themselves the germ [*fomitem*] of various changes in the weather or diseases just as an urn contains tokens, we will be able in just the same way to determine by observation how much more easily in these subjects this or that event may happen."[16]

When Einstein wittingly commented, "God does not play dice with the universe," he was referring to what was at that time the new quantum mechanics, which could not predict outcomes with certainty.[17] Fortune will never admit that the outcome of a rolling die is not really random, just as the lottery commission will never admit that the Ping-Pong balls that pick lottery numbers are not picking randomly. No one has yet devised a physical machine that gives absolute random numbers. "Rolling dice," writes physicist Robert Oerter, "is not inherently random; the outcome only seems random because of our ignorance of the little details, the hidden variable (like launch angle and friction) that determine the outcome of the role."[18] Most phenomena in our universe (especially those affected by atomic authorities) have far too many of those hidden variables for mathematics to predict their outcomes. We are generally ignorant of the minutiae of such wonders. Yet we have this marvelous gift that was a secret up until the end of the seventeenth century, which gives us a clue that the key to understanding randomness—as well as the means of predicting the future—is in the understanding that most happenings of the non–quantum mechanical world obey the weak law of large numbers, even though each individual event has no history of its past. Whether God plays dice or not, long-term trends of expectations are predictable and, almost always, assured.[19]

Bernoulli's proof rested on the number of ways objects can be combined, an accounting that has nothing to do with the random winds of Fortune. Edith Dudley Sylla, a distinguished translator of the *Ars Conjectandi*, tells us that Bernoulli explained the connection through theology. She wrote, "He convinces that in the mind or will of God there are distinct or determinate cases, known timelessly to God, that manifest themselves in experience or observation over time." The "timelessness" she refers to is Bernoulli's ignoring of time in random success ratios. Sylla points to Bernoulli's argument that "there is no real difference between throwing one die successively over time, on the one hand, and throwing all at once a number of dice equal to the times the one die was thrown."[20]

Expected Value

Expectation, measured by *expected value* (soon to be defined), is the harness that reins the mysteries of uncertainty. It, along with *standard deviation*, which measures of the amount of scattering from expectation, gives us a window into the stochastic (random) world. Those two measures—expected value and standard deviation—are the nuts and bolts of frequency distribution statistics, measurements of how closely data huddle around some central value. Miraculously, from them and simple algebra we have—if not direct management—at least a soft measure of phenomenological chance by way of the weak law of large numbers. In the physical world, every throw of a die and every descent of a Ping-Pong ball is influenced by a huge number of hardly measurable changing forces and circumstances (velocity, trajectory, air currents, gyro effects, angular momentum, impact, etc.), yet determinable in the ideal world of mathematics.

In 1657 Dutch mathematician and astronomer Christiaan Huygens published *De Ratiociniis in Aleae Ludo* (On Reasoning in Games of Chance), which for the next half-century remained the chief text on probability.[21] In it is the first printed recognition of the difference between number of successes and the *likelihood* of the number of successes.[22]

> Although the outcomes of games that are governed purely by lot are uncertain, the extent to which a person is closer to winning than to losing always has a

> determination. Thus, if a person undertakes to get a six on the first toss of a die, it is indeed uncertain whether he will succeed, but how much more likely he is to fail than succeed is definite and can be calculated.[23]

Huygens gives the example of a game of chance where you have to pay to play. A person hides three coins in one hand, seven in the other, and offers you the coins of whatever hand is chosen. You must pay to continue. But the question is: how much should you pay to play? Huygens's first proposition tells us the answer: "If I may expect either a or b and either could equally easily fall to my lot, then my expectation should be said to be worth $(a + b)/2$." The answer is 5, that is, the *expected value* (the amount you are expected to get in return), or the average of 3 and 7. It is not at all clear that Huygens understood the remarkable power his notion would have on the future of risk analysis, gambling, and science itself. But he did understand that the nucleus of the theory of probability is simply the expected value. It would have been wholly premature for a mid-seventeenth-century mathematician to know the real truth: that all of nature's random performance, including the behaviors of annuities, insurance, meteorology, and medicine, as well as games of chance, could be more or less foreseen by calculations of expected values. In general, expected value is calculated by multiplying the probability by the payout. In most cases it is the weighted average of all possible values that could occur, where the weighting is the probability. It is the sum of all possible values after each value is multiplied by the probability that it will occur. That makes sense; after all, you would *expect* to gain fifty cents on a dollar for each coin flip betting on tails.

For example, take the Texas Lotto lottery. Table 5.1 shows the results of matching 3, 4, 5, and 6 numbers. To get the expected value of playing the game, multiply the probability and payout of each possible match and take the sum over all matches.

Table 5.1.

Match	Payout	Probability
6 numbers	Jackpot	0.000000038
5 numbers	$2,000	0.00001115
4 numbers	$50	0.000654878
3 numbers	$3	0.013157894

If we suppose that the jackpot is worth, say, $2 million, then the expected value is 0.000000038($2,000,000) + 0.00001115($2,000) + 0.000654878($50) + 0.013157894($3) = $0.171517582. In other words, the actual value of a ticket to play is only seventeen cents.

At that early stage in the history of probability, people were using expected value as a measure of risk without knowing that it would turn out to be the most natural measurement of central tendency, the measurement of the tendency for data to cluster near some central value, as was seen in Figure 5.3.

Chapter 6

Long Strings of Heads

ACCORDING TO THE World Health Organization, the ratio of male births to total births in the entire world is 0.515.[1] When we look at specific areas or specific countries, those odds are very far from even. Mexico had a very low male-to-female ratio, whereas the United States and Canada had a higher than even ratio.[2] With a world population past 7 billion, however, the odds of a male birth vs. a female birth are close to even. The reason is simple. Human sperm has equal numbers of X and Y chromosomes that have an equal chance at conception. It is a fair coin flip.

After flipping a fair coin 7 billion times, we might expect that half of the flips ended with heads facing up. But should we expect to see a string of a million consecutive heads? A coin-tossing machine teaches us that in spite of the random interferences of a coin's trajectory, the coin could be made to land on heads 100 percent of the time.

The probability of a tossed fair coin coming up heads is 1/2. We know from mathematics that as the number of tosses grows, the ratio of heads to tails more and more likely approaches 1. Heuristic judgment muddles the meaning of that last sentence into a belief that somehow a long string of tails will be made up by a balancing string of heads. It's easy to fall for the erroneous impression that if a face has not come up for a very long time, the chances of its appearance increase with every turn, even though we know that theoretically, every time a coin is flipped, the odds for and against each outcome are precisely the same—the coin is just as likely to land on heads as tails. It's just that people tend to muddle the difference between outcomes and frequencies.

Long streams of heads might happen. I've seen very long streams of heads. It may seem intuitively odd that this should happen, but consider this: Suppose you toss a coin ten times and it comes up heads seven times. The proportion of heads to tails is then 7 to 3. Now, popular intuition suggests that for the next ten tosses, tails should appear more than six times to counterbalance the more than expected number of heads that have already appeared. But the coin has no memory of what it did before, only a history recorded by the person who is watching the result. There is nothing to prevent the coin from coming up heads for the next five hundred tosses, and yet we would be surprised if it did.

Figure 6.1 represents a computer-generated cumulative outcome of five hundred repeated coin flips (+1 for each head and –1 for each tail). The horizontal line represents 0. Heads and tails alternate their leads. It is a horse race between two horses of equal odds. This is what you might expect. Normal intuitive judgment favors the opinion that penny-flip graphs should bounce over and below the zero line. However, most often such graphs favor one side over the other for long periods of time.

Absolute randomness as a theory is not the same as absolute randomness in the real, physical world. Those numbered Ping-Pong balls that first whirl around in those acrylic balloons to determine a lottery number are not escaping to a channel randomly, though for the casual observer they are certainly delivering unpredictable numbers. The coin toss that determines a kickoff in American football is quite far from being random. In fact, the result of a coin toss is simply a matter of physics. Machines have been built that can toss a coin any number of times—a thousand, a million—to show that every time, the coin will land on heads.

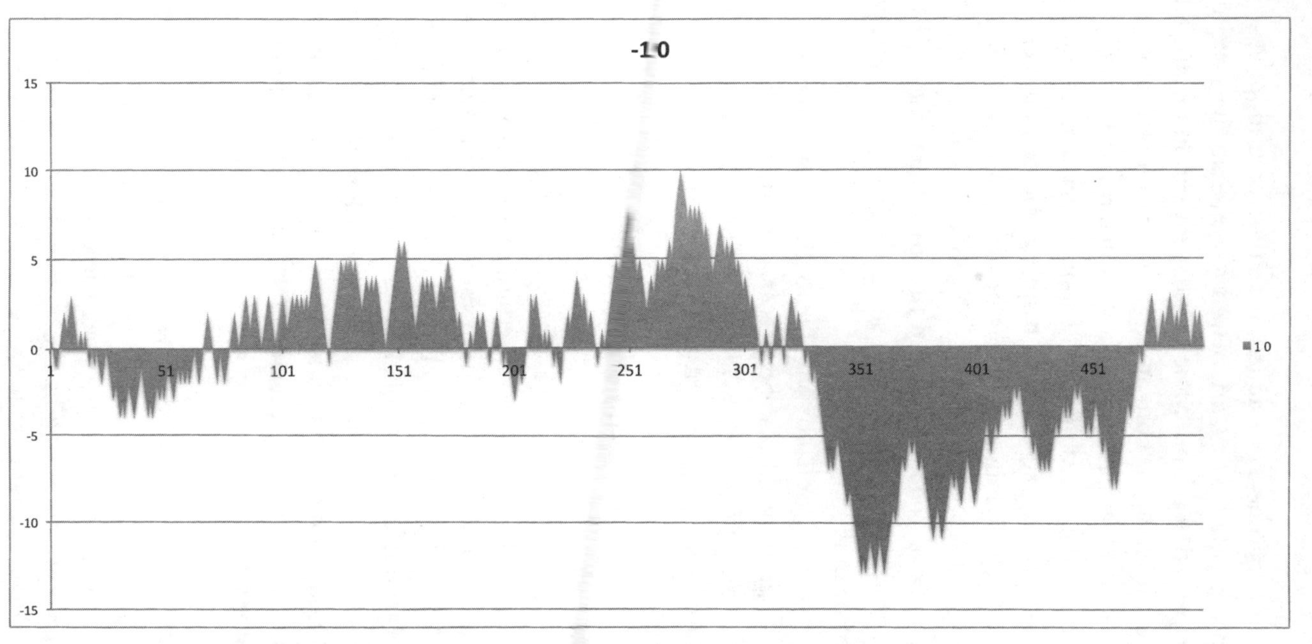

Figure 6.1. Cumulative frequencies of heads vs. tails as a function of the number of tosses.

Recent experiments designed to analyze coin flipping show that coins, even fair coins, are biased to come up the same way they start, and that the outcomes are dependent on the angle between the normal to the coin and the angular momentum vector. In other words, the flight of the coin is determined by its initial conditions. Diaconis, Holmes, and Montgomery built a coin-tossing machine that tossed coins by a spring-released ratchet.[3] For that machine, any coin that starts heads up always (100% of the time) lands heads up. So, the result of a coin toss is physics, not randomness. The human hand that tosses and the manifold variables in the environment cause variations in outcome that appear to be random.

But we can be fooled by the illusion that the coin is actually flipping when it could just be precessing through the air like a slowly rotating gyroscope. The coin's orientation in flight is dictated by its angular momentum vector, which might always be pointing up. So, a coin that starts heads up might be always heads up as it follows its trajectory, giving the appearance that heads and tails are spinning.

When it comes to those real coin flips where the outcomes are determined by the slightest interference from ground tremors of a thousand miles away or that meddlesome chaos-causing butterfly in the Pacific, things are different. But *different* means neither reasonable nor fathomable. Its landings might very well be random, but our human perception of randomness is often at odds with our own premonitions of random outcomes. Since the coin has no memory of its previous outcomes, we should not be surprised if it were to come up heads a hundred times in a row, but we are.

Figure 6.2 tells a strange story. The outcomes go as expected until around the 45th flip, when tails take over and become "hot" for the next roughly 105 flips! Then, there is a reasonable period where heads are hot, bringing the cumulative value closer to 0. But then again, at around flip 286, tails take over and stay in

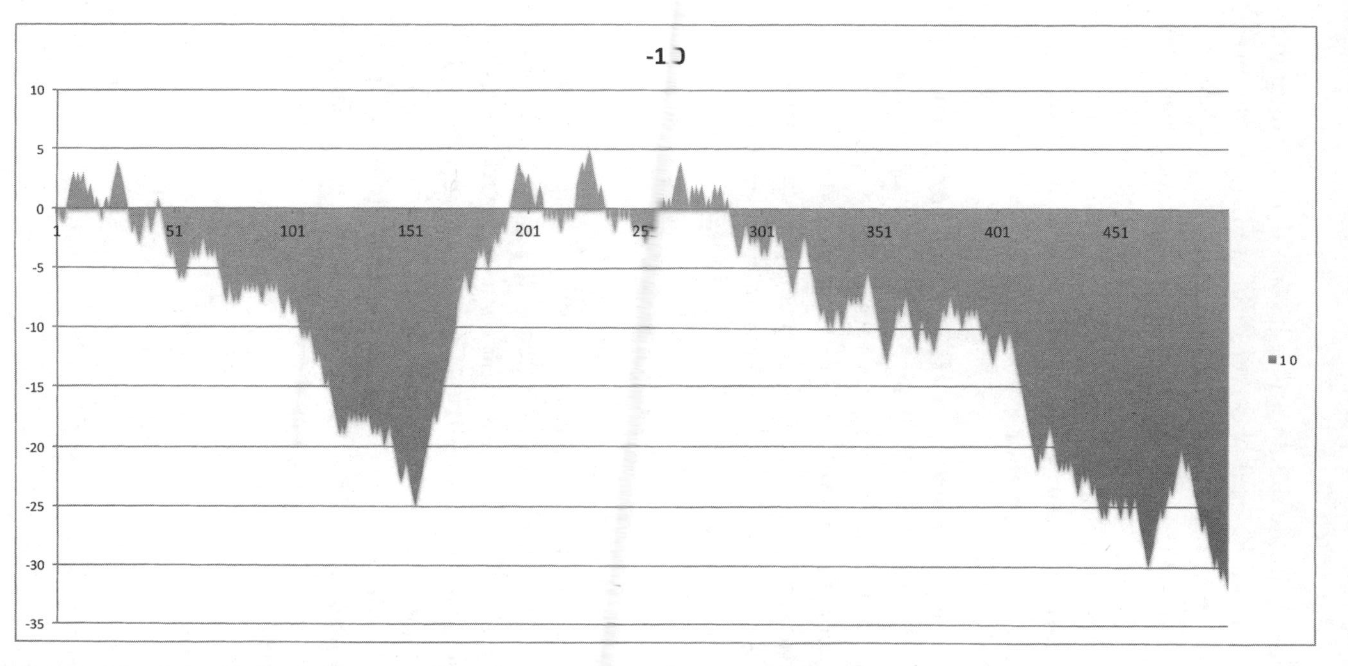

Figure 6.2. Cumulative frequency of heads vs. tails as a function of the number of tosses.

the lead once again for the duration. It's not that our intuition about what should happen was ever disobeyed. The actual ratio of heads to tails will surely get closer and closer to 1 over a much larger range that never materialized, but we don't see that happening in the short term. In 500 flips, tails appeared just 12 times more often than heads. That does seem rather close, but often the sequences of tails vs. heads can make a large difference in the cumulative results. For example, consider the next trial shown in Figure 6.3.

Heads are in complete control. The cumulative outcome puts heads in the lead for almost the entire duration of flips giving the impression that tails will never gain a lead.

The results of a million coin flips are broken down in Table 6.1, the virtual results of a computer-generated run of 1 million flips. The ratio k/N, where k represents the number of successes and N the number of trials, is called the *observed success ratio*. The column on the right side of Table 6.1 lists the absolute values of the difference between the observed success ratio and 1/2 the mathematically predicted success ratio.

The weak law of large numbers does not preclude any of the unlikely events from happening often in the game, early on or later. In fact, even if the success ratio comes close to the mathematically predicted success ratio, there is no assurance that it will continue to stay close. A slightly stronger mathematical result tells us that, although the success ratio is likely to converge on what is mathematically predicted, the actual success values tend to behave increasingly wildly as the number of events increase. It's counterintuitive, yet true.

The weak law of large numbers applied to any event whose probability of success is p tells us that the probability that $\left|\frac{k}{N} - p\right| < \varepsilon$ comes closer to 1 as N gets larger. Take $\varepsilon = 0.0001$ (this is arbitrary) with $p = 1/2$ for coin tossing, and ask how likely $\left|\frac{k}{N} - \frac{1}{2}\right|$ is less than 0.0001. Notice (Table 6.1) that $\left|\frac{k}{N} - \frac{1}{2}\right|$ jumps

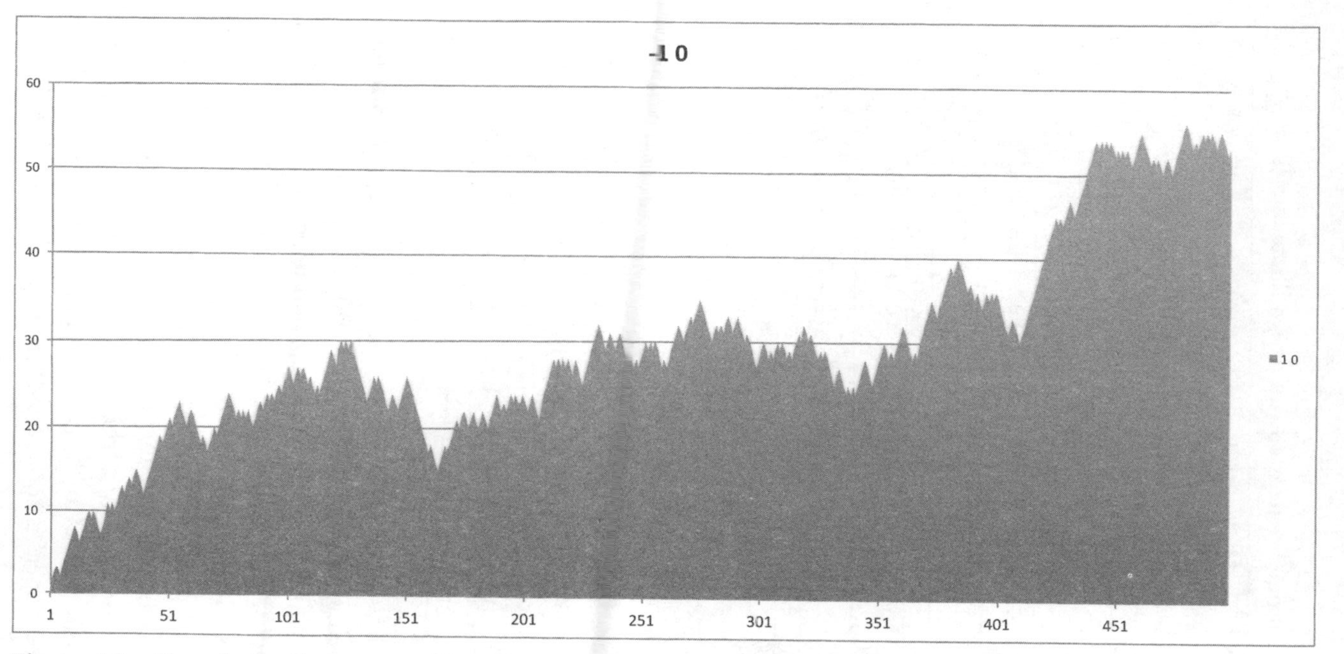

Figure 6.3. Cumulative frequency of heads vs. tails as a function of the number of tosses.

Table 6.1. Computer-generated run of 1 million coin flips

N=number of flips	*k=number of observed heads*	*k/N*	$\lvert k/N - \frac{1}{2} \rvert$
2,500	1,254	0.5016	0.0016
5,000	2,561	0.5122	0.0112
7,500	3,758	0.5012	0.0012
10,000	5,018	0.5018	0.0018
12,500	6,226	0.4981	0.0019
15,000	7,411	0.4941	0.0059
17,500	8,805	0.5031	0.0031
20,000	10,007	0.5004	0.0004
100,000	49,995	0.49995	0.00005
200,000	99,883	0.49942	0.000585
300,000	150,279	0.50093	0.00093
400,000	200,186	0.500465	0.000465
500,000	250,007	0.500014	0.000014
600,000	300,342	0.50057	0.00057
700,000	349,788	0.499697	0.000303
800,000	400,257	0.50032125	0.00032125
900,000	449,688	0.49965333	0.0034667
1,000,000	500,010	0.50001	0.00001

Table 6.2. Detail from Table 6.1

N	*k = Heads*	*Tails*	*Heads-Tails*	*(H - T)/N*	*|k/N-1/2|*
5,000	2,561	2,439	122	0.0244	0.0122
67,500	33,371	34,129	-758	-0.01122963	0.005614815
82,500	41,597	40,903	694	0.008412121	0.004206061

around for low values of *N*. But it seems to be jumping for high values as well. From 100,000 to 200,000, it increases. Even from 800,000 to 900,000, it increases before it decreases at a million. The misleading impression is that the difference between heads and tails should approach zero. But it says nothing about the volatility in approaching through high numbers. As we can see, that volatility increases as the number of flips increase.

So, what is going on here? It seems that higher *N* has some freedom from the law of large numbers because in vastness of large numbers there is room for more unnoticeable error.

For 5,000 tosses there were 2,561 heads and 2,439 tails with a difference of 122. That gives a 2.4 percent error, which does not seem so bad. But not knowing the distribution of those heads, it may be that 122 heads were thrown consecutively. Taking that view, imagine 758 tails thrown consecutively in 67,500 tosses, or 694 heads thrown consecutively in 82,500 tosses. In other words, there is no mathematical law that precludes the possibility of an enormous number of heads being thrown consecutively when *N* is large.

Chapter 7

Pascal's Triangle

IN THE PHYSICAL world there is no such thing as perfect symmetry, human-made machines of infinitesimal tolerance, or ideal models. It is a profoundly entangled hidden-variable world whose happenings are too hard to pin down by exact measure. So, real chance happens, and we often resort to probabilistic pictures to understand the more perplexing phenomena of chance.

What if you were to have the misfortune of having the rare disease myelodysplastic syndrome, a cancer in which the bone marrow does not make enough healthy blood cells? You would be faced the dilemma of whether to accept a bone marrow transplant with a 70 percent chance of success, or to do nothing with a 70 percent chance of dying within the next ten years. Of course, the transplant has its risks. Between needing chemotherapy and the risks of infection, there would be about a 30 percent chance that you would die within the next six months.

Brian Zikmund-Fisher, who teaches risk and probability at the University of Michigan School of Public Health, faced just such a dilemma back in 1998. Diagnosed with myelodysplastic syndrome, he was told that without treatment he would have only ten years to live, and with treatment he would have a 70 percent chance of having a normal life.[1] He gambled on the transplant. The point here is that the odds say nothing about the individual. The 70 percent chance comes from statistical data collected over hundreds (perhaps thousands) of individuals who found themselves in his dilemma—a national, nonlocal statistic. Statistical groupings are about trends and possibilities, not about individual cases that could win or lose.

Take some event that you might consider as rare. Its mathematical chances may even be one in a million, but that might be because it is being evaluated as a local phenomenon. An example might be a squirrel hit by lightning while it was crossing a street. When we talk in that familiar language of chances, we often speak figuratively, without any systematic method of backing up our terms. So, the one in a million is generally applied to what we think happens in a somewhat wide area of the United States. But the United States is a vast country. We see that by just flying over it and seeing all those tiny houses and tiny trees and vast acres of green. We think neither about how many squirrels are out there, nor about how many are crossing streets at any one time. Scientists estimate that there are 1.12 billion squirrels in the United States, three times the human population. And squirrels are always crossing roads.

With 1.12 billion squirrels, 4.09 million miles of roads, and 3,794,101 square miles of land in the United States, it is plausible that at any one minute of any day, there are on average 300 squirrels crossing US roads.[2] During thunderstorms, there might even be more. On average, each year there are more than 110,000 thunderstorms in the United States. There are many more thunderstorms in summer than in winter, which makes the likelihood of a lightning bolt directly hitting a squirrel in summer very great indeed.

Every event in nature has to account for a vast number of indeterminate possibilities. The toss of a die may strongly depend on its initial position in the hand that throws it and more weakly depend on sound waves of a voice in the room. Those are just two external modifiers that guide the die to its resting position.

How it strikes the table, the precision of its balance, how it rolls off the hand, and the elasticity of its collision with the table will influence which side faces up when it comes to rest.

Consider playing a game where there are only wins and losses with no tie score. Let X represents the outcome of an event, and $P(X)$ the probability that the event actually happens. If you were flipping a coin, for instance, $P(heads)$ would equal 1/2, and so would $P(tails)$. In American roulette, there are thirty-eight pockets on the wheel, including 0 and 00: eighteen are red; eighteen, black; 0 and 00 are green. If you were betting on red, $P(red)$ would equal 18/38 or, more simply, 9/19, and $P(nonred)$ would be 10/19. If you were rolling a die, hoping for an ace (a 1), $P(1)$ would equal 1/6.

Pick any such game and play it four times and ask: what is the probability of winning zero, one, two, three, or four times? It's a fair question, since real gambling involves cumulative strings of wins and losses. Think back to Joan Ginther's four lottery wins. You might also want to know the odds of doing better than breaking even, or at least the odds of not losing more than twice in four bets.

Let strings of *W*'s and *L*'s represent respective strings of wins and losses. Losing all four times would be marked by *LLLL* and winning all four times by *WWWW*. There is only one way to win all four times and only one way to never win. What about winning once in all four rounds? There are four ways of winning once in all four rounds, represented as *WLLL*, *LWLL*, *LLWL*, and *LLLW*. And, of course, there are four ways of losing once in all four rounds. What about winning twice in all four rounds? Winning twice would be represented by the six configurations *WWLL*, *WLWL*, *WLLW*, *LWWL*, *LWLW*, and *LLWW*. We have not accounted for the order of wins and losses, since we listed them in strings of four letters without regard to order. In mutually exclusive events where the outcome of one event has no memory of any other, such as rounds of roulette or coin flipping, the probabilities of one *or* the other of two things happening is the product of the probabilities of each. From what we said in Chapter 4, if A and B are possible outcomes, the probability of A *and* B happening is the product $P(A)P(B)$, and the probability of A *or* B happening is the sum $P(A) + P(B)$.

Now, let's take the middle case of having two wins in four rounds. To simplify the notation, we shall let p represent $P(W)$ and q represent $P(L)$. The probability of one single win is p, and since wins and losses are mutually exclusive (i.e., each round does not depend on the round before it), we see that the probability of having two wins in four rounds is p^2q^2. This is because you would have to win twice *and* lose twice, and when the logical connective is *and*, the probabilities are multiplied. But, as we have seen, this can happen in the following six distinct ways: *WWLL*, *WLWL*, *WLLW*, *LWWL*, *LWLW*, and *LLWW*.

Since *or* is the logical connective, the probability of any one of these events happening is $ppqq + pqpq + pqqp + qppq + qpqp + qqpp$, or simply $6p^2q^2$.

Table 7.1, constructed from knowing the values of p and q for the three different games, shows the probabilities of winning zero, one, two, three, or four times in four rounds.

In theory, for both roulette and coin flipping, according to Table 7.1, a player is most likely to win twice in four rounds. We could construct a table of probabilities for a hundred rounds of roulette and coin flipping, though it would be dauntingly long and impractical. Instead, let me just say that in a hundred rounds of calling heads on the flip of a coin, a player would be most likely to win fifty times, but in a hundred rounds of playing red in roulette he or she would be most likely to win (as we shall see) only forty-seven times.[3] The gamblers' holy grail is to know which forty-seven.

Note the symmetry with both roulette and coin flipping, the asymmetry with dice, and extreme asymmetry with lotteries. What about the column for roulette in Table 7.1? Pictured as a

Table 7.1.

Number of wins	Number of ways a win can occur	Probability of winning	Probability of red in roulette	Probability of heads in coin flipping	Probability of tossing 7 with a pair of dice	Probability of winning Texas Lotto lottery (approximately)
0	1	$1q^4$	0.077	0.0625	0.4823	0.999999848
1	4	$4p^1q^3$	0.276	0.25	0.3858	1.52×10^{-7}
2	6	$6p^2q^2$	0.373	0.375	0.1157	8.66×10^{-15}
3	4	$4p^3q^1$	0.224	0.25	0.0154	2.19×10^{-22}
4	1	$1p^4$	0.050	0.0625	0.0008	2.09×10^{-30}

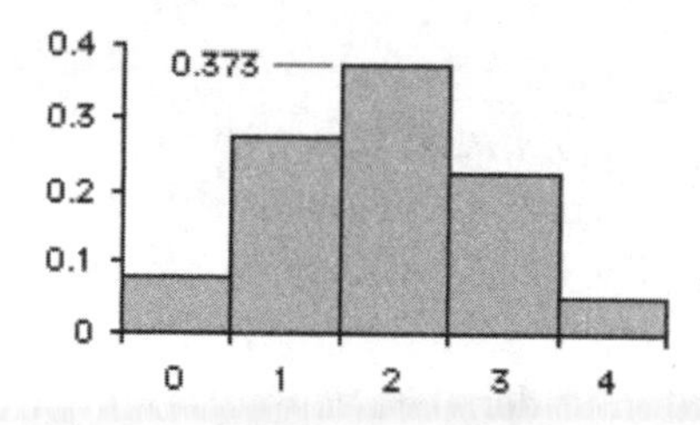

Figure 7.1A. Probability of winning on red in four rounds of roulette.

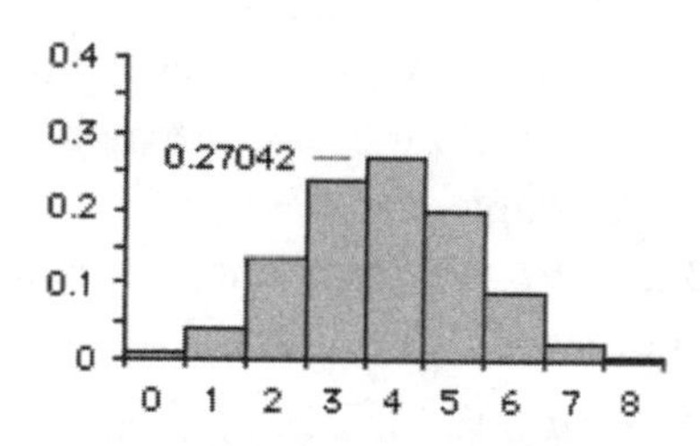

Figure 7.1B. Probability of winning on red in eight rounds of roulette.

bar graph plotting the number of times red appears vs. the probability of getting that number of reds (see Figure 7.1A), there is a skewed symmetry about the number 2, while the center of gravity (the geometric balancing point) seems to be over a number just less than 2. When the number of rounds increases to 8, the skew is even more pronounced (see Figure 7.1B).[4]

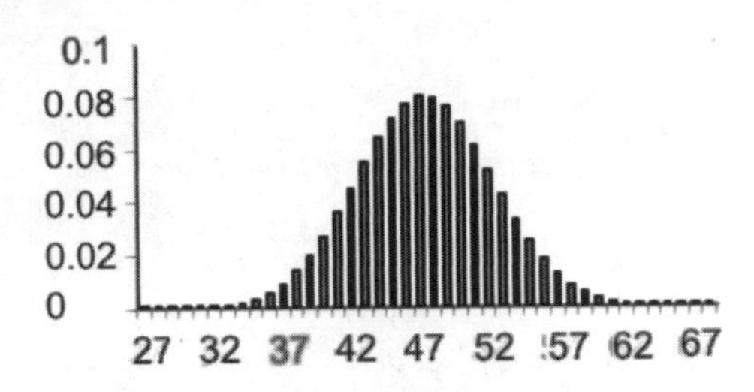

Figure 7.2. Probability of winning on red in one hundred rounds of roulette.

Increasing the number of roulette rounds smoothens out the graph. For one hundred rounds there would then be 101 rectangles, each having a base one unit wide.[5]

Figure 7.2 is what is called a *frequency distribution*. The height at each number of successes tells us how frequently those successes are expected to occur. The bars are distributed over the horizontal axis in such a way that the total sum of their areas equals 1. In other words, the area under the graph accounts for 100 percent of all possible events. Most of the frequency distribution is concentrated between 32 and 62, with the highest bar at 47. Below 32 and above 62, the probabilities are so small we cannot see them on the graph. For example, $P(31) = 0.00034$ and $P(63) = 0.0006$. Red is far less likely to turn up twenty or eighty times yet, like all coincidences, not impossible.

For coin flipping, where p equals q, there is perfect symmetry. But p does not have to equal q. We find a skew symmetry that gets more pronounced the farther p is from q. In Table 7.1 we see perfect symmetry in the fifth column from the left and hardly any symmetry in the seventh column. And yet all the calculations are coming from the third column, a result of the magnificent so-called *Pascal triangle*, a key to the warehouse of probability tools.

Pascal's triangle is the following triangular arrangement of numbers:

Figure 7.3. Pascal's triangle.

Each number in Figure 7.3 is the sum of the two numbers directly on the line above it; for example, the third number (10) on the fifth line from the top is the sum of the 4 and 6 on the fourth line. First notice the symmetry, and then that the numbers are the same numbers that we see when we expand the powers of a sum of two variables, say, p and q. We find these same numbers when we expand the power $(p + q)^n$. For example, when $n = 2$, $(p + q)^2 = (p + q)(p + q) = p(p + q) + q(p + q) = p^2 + pq + qp + q^2 = p^2 + 2p^1q^1 + q^2$.

If we perform and list the expansion for $n = 1, 2, 3, 4, 5, 6, \ldots$ we get the following triangular-looking array:

$$(p+q)^0 = 1$$

$$(p+q)^1 = 1p^1q^0 + 1p^0q^1$$

$$(p+q)^2 = 1p^2q^0 + 2p^1q^1 + 1p^0q^2$$

$$(p+q)^3 = 1p^3q^0 + 3p^2q^1 + 3p^1q^2 + 1p^0q^3$$

$$(p+q)^4 = 1p^4q^0 + 4p^3q^1 + 6p^2q^2 + 4p^1q^3 + 1p^0q^4$$

$$(p+q)^5 = 1p^5q^0 + 5p^4q^1 + 10p^3q^2 + 10p^2q^3 + 5p^1q^4 + p^0q^5$$

$$(p+q)^6 = 1p^6q^0 + 6p^5q^1 + 15p^4q^2 + 20p^3q^3 + 15p^2q^4 + 6p^1q^5 + p^0q^6$$

For any n, the constants in the expansions of the binomials $(p + q)^n$ are exactly the numbers in Pascal's triangle.

This triangle has a history that starts long before Blaise Pascal.[6] It appeared in the works of twelfth-century Chinese algebraist Chu Shï-kié and later appeared on the title page of Petrus Apianus's *The Arithmetic Book* in 1527 (which appears in the painting *The Ambassadors* [1533] by Hans Holbein the Younger) more than a century before Pascal ever investigated the triangle named after him.[7] In modern Iran the triangle is known as the *Khayyám triangle*, after the famous Persian poet-mathematician Omar Khayyám who used that triangle in the twelfth century to compose a method for finding nth roots. In modern China it is called *Yang Hui's triangle* in honor of another mathematician who introduced it to China in the thirteenth century. In Italy it is *Tartaglia's triangle*, after mathematician Niccolò Tartaglia, who lived a century before Pascal. However, Pascal, a collector of many results that had already been known about the triangle, used them in probability theory.[8]

Probability Distributions

Figure 7.2 shows the probability of winning on red in one hundred rounds of roulette. We have seen how the graph shapes itself from the calculation examples in Table 7.1 and the coefficients coming from the binomials $(p + q)^n$. The distribution of bars in the graph is rightly called a *binomial distribution*. The word *binomial* comes from the construction based on the two monomials p and q. As we increase n, the tops of the bar graph smoothen out to look more like a bell-shaped curve. The larger n is, the smoother the curve.

Pick some large n. We will transform the bar graph while preserving its area, and hence the probability. Since the base of each bar is one unit in width, the distributions of probabilities are represented by the areas of the rectangles, as well as by the heights. Modifications, through clever shifting, shrinking, and magnification, bring us a new graph that preserves all the useful information of the original.[9] Of course, now, in the modified graph, the vertical axis will no longer represent probability. That job rests with the areas of the rectangles, and those areas have not changed because we magnified the vertical and contracted the horizontal by the same factor.

What have we achieved? Here's the marvel, an inspired idea. The curve (the binomial distribution bar graph that appears in Figure 7.2) that represents the probability of winning on red in one hundred rounds of roulette, may be closely approximated by one particular mathematical curve. The important thing to understand here is that this one particular curve describes a great many natural phenomena resulting from chance behavior. Astoundingly, this one particular curve models events of roulette, though it has no apparent connection to balls falling into red pockets of roulette wheels. More surprising yet, that same curve models coin flipping just as well. Just one curve models the probabilities of so many different occurrences. To get information about the probability of a particular occurrence, we must feed some information back into the model. We must provide two numbers—the mean (average) and standard deviation (measure of the spread from the mean).[10] Those two numbers give the model information about, say, roulette specifically, namely that the probability of success p (the ball falling into a red pocket) is 9/19. Once we have that specific p and N (the number of rounds of roulette played), we are able to compute the standard deviation for our particular game of playing red in roulette.[11] It is a measure of how scattered the outcomes are from the mean, the *standard deviation from the mean*, or more commonly known as the *standard deviation*.[12]

So, every binomial frequency curve is transformed by a math trick (through shifting and scaling) into the special and powerful *standard normal curve* whose graph is pictured in Figure 7.4.[13]

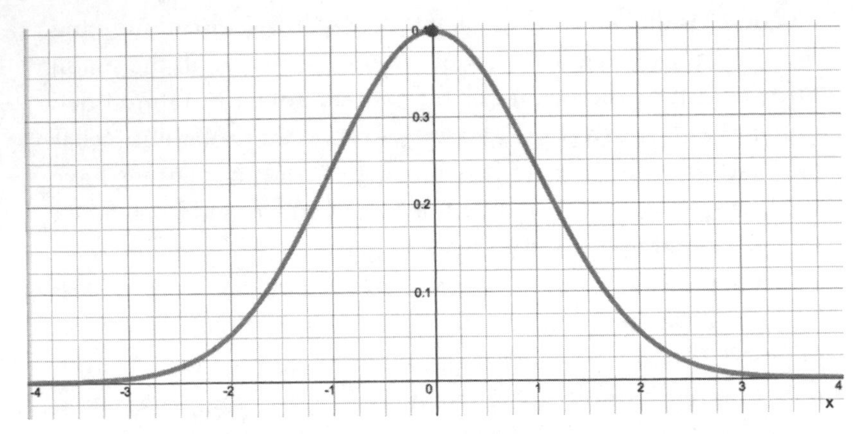

Figure 7.4. The graph of the standard normal curve.

The numbers at the base of the curve in Figure 7.4 are counting the numbers of standard deviations from the mean. We have bundled the trials into groups of standard deviations. Individual probabilities of event outcomes are no longer visible. The variable *X* under the curve in Figure 7.4 measures the deviation of the number of successes away from the likeliest number of successes. So, *X*, the horizontal unit, is measured in standard deviations. The height of the curve is no longer a probability, because it has been scaled and shrunk to preserve the area under the curve. But we get several very valuable bits of information in return for all that scaling and shrinking. One is that about 68 percent of area under the curve rests over one standard deviation of the mean, and that about 95 percent of the area rests over two standard deviations of the mean. Another is knowing that one standard deviation is marked by the inflection points, those points on the curve where the shape of the curve changes from being concave down to concave up.

Although one standard deviation for a red outcome on a hundred rounds of roulette is not the same as one standard deviation for heads on a hundred rounds of coin flips, the curve in either case is magically the same. The interpretations of what those curves mean will be different. And although the curve in Figure 7.4 may be the same for many different gambling distributions of odds, the axis markings must be interpreted by specific calculations of the mean and standard deviation. That information will depend on the number of rounds, and probabilities of positive outcomes, specific to the game.

When we examine a frequency distribution, we tend to look mostly at the deviation from the normal. But what happens far outside the normal range can have a devastating effect on the overall cumulative outcome. We pay little attention to that outside area, because we are mostly thinking of questions of central tendency and events that are highly likely, not about what could happen in the most unlikely cases.

Do we ever take into account unlikely cases or worst-case scenarios? Or do we simply say that they are so rare that we should just dismiss them? They are the coincidences or the flukes of nature, the actual physical events that ride on the tailwinds of chance. As the number of flips of a fair coin continues to grow, the total number of heads may far exceed the total number of tails (or vice versa). For instance, if you flip a coin one hundred times, heads could come up every time—unlikely but possible, despite the odds being 1 to 1. To be more conservative for a moment, consider the case where in one hundred throws the events show 41 heads and 59 tails, or probabilities of 0.41 and 0.59, respectively. Sounds like a large difference, but in one hundred throws, the actual difference between heads and tails is just 18. However, if you flip the coin five hundred times (as we have done in Chapter 6) and find that the ratios have shrunk much closer to a probability of 1/2—say, to where the ratios of heads to total tosses equals 0.45 and tails to total tosses is 0.55—there will be 225 heads and 275 tails, a difference of 50.

In other words, the differences can keep growing even when the ratios are shrinking to 1/2. Top this off with the understanding that there is no prediction for the distribution of the results, we find that as the number of flips grows, so does the possibility of larger and larger numbers of repeated heads. We could have flipped the coin one hundred times, taken a break, flipped it another one hundred times, and continued that way. Each time could have counted as a new time. So, how is it that the difference between tails and heads could be 50 in five hundred flips and possibly only 10 in one hundred flips? When does the 50 occur? Couldn't it happen during the last hundred flips, all in a row? Sure, that too would be a coincidence, but any possibility opens a modest chance!

In theory, roulette is played with an ideally spherical ball rolling and bouncing round a flawlessly balanced wheel with precisely spaced pockets in a perfectly steady room in some world we have never seen, in a world that has never existed. Real wagering happens in a physical world where balls and wheels are machined and manufactured to extremely severe tolerances, but human-made machines manufacture those balls and wheels. The connection between the ideal and physical is magical, yet so deeply tangled that we are dazzled by what we do not understand.

Ideal World vs. Physical World

In the physical world, we could test bona fide roulette wheels for fairness or biases by making a table of observations that could be pictured by a frequency distribution graph. Such a picture may not look anything like the graph of our perfect model, but if the wheel were indeed somewhat fair, and if we were to observe enough rounds, then the graph of observed outcomes should resemble (in shape at least) the graph in Figure 7.4. If we perform n trials of an experiment, we have n observed outcomes O_1, O_2, O_3, . . . O^n with respective probabilities p_1, p_2, p_3, . . . p^n to give us an observed probability distribution. For instance, as we noted earlier, in tossing dice, any one of six faces can be an outcome, each with a probability of 1/6. In a fair game, the experimental version of the distribution should turn out to closely resemble the theoretical distribution with the recognition that some discrepancies are bound to happen in a world that is not perfect.

In this context, *perfect* translates to *mathematical*. Understanding real odds comes from comparing data collected by observations with computations that are *expected* in a perfect world. Gamblers might know the odds are against them, and yet hope that the physical world might wander from its expectations to favor their bet. It comes from the powerful thought that *someone has to win*. They will risk heavily against the mathematical expectations of Fortune.

By analyzing the published records during a four-week period from July to August 1892 at a Monte Carlo casino, English mathematician Karl Pearson found that the mechanism, as machine precise and as perfectly adjusted for the roulette table as it could be, was not fully obeying the laws of chance.[14] Assuming mathematical precision, those laws tell us that a ball should be equally likely to fall in any one of the thirty-seven pockets of the wheel as any other.

Excluding the 0 pocket, there is an equal mathematical chance for the ball to fall into a red or black pocket.[15] This should mean that for a large number of physical spins, the ball should fall into the red pocket 50 percent of the time.

However, after spending a fortnight examining 4,274 spins of a Monte Carlo roulette wheel, Pearson found that their standard deviations from the normal were almost ten times what they were expected to be. The odds against such a thing happening with a fair roulette wheel are more than 10 trillion to one! Pearson wrote, "If Monte Carlo roulette had gone on since the beginning of geological time on this earth, we should not have expected such an occurrence as this fortnight's play to have occurred *once* on the supposition that the game is one of chance."[16]

By some miraculous coincidence, Pearson hit on one event so improbable that it could only occur once in the history of the world. Should that be a reason to doubt the fairness of the roulette wheel? A student of his tried the experiment again for another fortnight and found results less improbable yet expected to occur just once in five thousand years of continuous round-the-clock playing. Another investigator observed 7,976 spins during a fortnight at Monte Carlo, and computed the odds against a fair wheel as 263,000 to 1. Other experiments found the same coincidences. An 1893 observation of 30,575 spins showed odds of more than 50 million to 1. According to Pearson, "Monte Carlo roulette, if judged by returns which are published without apparently being repudiated by the Société, is, if the laws of chance rule, from the standpoint of exact science the most prodigious miracle of the nineteenth century. . . . "[17]

Divergence of theory from practice was so improbable that Pearson wrote, "*The odds are a thousand millions to one against such a deviation. . . .* "[18] His observations differed from the mathematically expected theory by odds of 1,000 million to 1 against! Eminent mathematician Warren Weaver wrote about a time in the 1950s when a wheel at Monte Carlo came up *even* twenty-eight times in straight succession. The odds of that happening are 268,435,456 to 1. Based on the number of coups per day at Monte Carlo, such an event is likely to happen only once in five hundred years.[19] And gaming expert John Scarne wrote about an occasion on July 9, 1959, at El San Juan Hotel in Puerto Rico, when a roulette ball landed on 10 six times in row. The odds of that happening are 133,448,704 to 1.[20]

If the game is expected to be fair and if what we observe is highly unlikely, then the game may not really be fair; however, we also know from the weak law of large numbers that extremely rare events have a reasonably high likelihood of happening at least once if the number of trials is sufficiently large.

Remember the famous *Casablanca* coincidence? It, too, is one that is so improbable that it could only occur once in the history of the world. In the movie, Rick Blaine, the owner of the nightclub Rick's, tries to save a young Bulgarian girl's fiancé, Jan, from losing all his money for an exit visa at the roulette table. Young, pretty, and naive Annina had asked Rick about the honesty of police captain Louie Renault, who promised her an exit visa for certain compromises.

Let's recall the next scene in the gaming room of Rick's Café. Jan is seated at the roulette table. He has only three chips left. Rick enters and stands behind Jan.

CROUPIER (*to Jan*): Do you wish to place another bet, sir?
JAN: No, no, I guess not.
RICK (*to Jan*): Have you tried twenty-two tonight? (*Looks at the croupier.*) I said, twenty-two.
(*Jan looks at Rick, then at the chips in his hand. He pauses, then puts the chips on twenty-two. Rick and the croupier exchange looks. The wheel is spun. Carl is watching.*)
CROUPIER: Vingt-deux, noir, vingt-deux. (*He pushes a pile of chips onto twenty-two.*)
RICK: Leave it there.
(*Jan hesitates, but leaves the pile. The wheel spins. It stops.*)
CROUPIER: Vingt-deux, noir. (*He pushes another pile of chips toward Jan.*)
RICK (*to Jan.*): Cash it in and don't come back.
(*Jan rises to go to the cashier.*)
A CUSTOMER (*to Carl*): Say, are you sure this place is honest?
CARL (*excitedly, in his lovable Yiddish accent*): Honest? As honest as the day is long!

The odds of a roulette ball falling into pocket number 22 twice in a row are 1,369 to 1, nowhere near the questionability we imagine when watching the film. It's fiction. Fair enough. In real life, in a fair game with those odds, we should not be surprised to see 22 be the winning number twice in a row. But Rick called it, and it happened exactly when he called it. That makes the odds against it far greater than 1,369 to 1.

Prior to that wonderfully fictional honesty at Rick's Café was the underhanded fictitious account of Signor Emanuel Ravelli (Chico) and The Professor (Harpo) playing bridge in the Marx Brothers movie *Animal Crackers*. Ravelli and The Professor (forever partners in crime) were drawing cards to determine partnerships in a game of bridge. Ravelli draws his card and announces that he's got an ace of spades. The Professor then draws and shows his card, leading Signor Ravelli's quip, "He's got ace of spades. Ha, ha! That's what you call coincidence."

Chapter 8

The Problem with Monkeys

WE ARE OFTEN DECEIVED by the magnitude of our world. It is bigger than we think; it is smaller than we think. A hundred years ago we stayed close to our towns and villages. My great uncles and great aunts in Poland surely didn't wander far from their shtetl. Today, because of our international mobility, we bump into friends and relatives without surprise. We don't quite fathom the hugeness of the world when we can get from New York to Hong Kong in fifteen hours. If I ask you how many people in the world have committed suicide in the time it took you to read this paragraph, you might very well say zero. But to give you some sense of how large this world really is, let me tell you that, according to estimates from the World Health Organization, on average every forty seconds someone, someplace in the world, performs a successful suicide. That's 2,160 people, every day, on average! The rate varies according to country. In India, where suicide is illegal, the rate is almost double the global average.

By definition, coincidences are events that happen without apparent cause. Apparent to whom? It does not mean there is no cause. The world generally works by cause and effect. I say generally, because there are acausal phenomena in physics, psychology, and religion. But the word *apparent* tells us that the moment we learn the cause of a coincidental phenomenon, its status diminishes to a simple time-space event. That must mean that coincidences are relative to the people affected by them. It also means that the unapparent cause is there, waiting to be discovered. If there is no cause at all, then it happens by chance.

The odds of drawing an ace of spades from an ordinary well-shuffled deck of fifty-two cards is 51 to 1 against, meaning that there are 51 ways of not drawing that card and 1 way of drawing it. The chances of drawing an ace of any suit are 12 to 1 against. It simply means that by drawing thirteen cards, you have a pretty good chance of picking an ace. What actually happens is a matter of chance.

Suppose you drew that ace of spades, put it back in the deck, and drew again. Your odds of drawing that same card are still 51 to 1, even though they were 2,703 to 1 to have drawn it twice in succession. That is, in drawing the ace of spades again, two things had to happen, each with 51 to 1 odds, so the probability of drawing that ace twice is $(1/52)(1/52) = 1/2704$, and therefore the odds of drawing it twice are 2,703 to 1. It may seem to be paradoxical, since the second draw should be no more challenging than the first.

Even with that poor chance, it is still possible to draw that ace of spades a second time. By experience, we know that happens quite frequently. You might bet a dollar that you can draw that ace of spades twice in a row, but don't bet the farm. The smart thing to do is bet that dollar with a payback no less than 2,703 to 1 that you would draw the ace of spades again. That way, if you have a few thousand dollars to spare, you could play the game a few thousand times and come out . . . ha-ha . . . with a pretty reasonable chance of winning at least once.

Of course, it is far more unlikely to draw the ace of spades a third time in a row, or a fourth time. The probability of drawing it a fourth time is $(1/52)(1/52)(1/52)(1/52) = 1/7{,}311{,}616$, so the odds against it are 7,311,615 to 1. Unlikely, but not impossible. This time don't even bet a dollar. Indeed, it is not impossible to draw that same ace fifty times in a row, or a hundred times in a row, or any other great number of times.

If you did draw that ace of spades four consecutive times, you might become suspicious of the deck. But chance is a funny thing. Nothing in the laws of chance stops that ace of spades from popping up four times consecutively. No more so than throwing musical notes in the air and having them land to form a Beethoven sonata. You wouldn't bet that you could write music like Beethoven by throwing notes in the air. But it is certainly possible that by throwing notes in the air often enough, some sort of reasonable sonata might come out.

Let's now suppose you are playing poker with ten other players. The odds of drawing a royal flush of clubs, say, A♣ K♣ Q♣ J♣ 10♣, are 2,598,959 to 1. Why? Because there are 52 distinct ways of being dealt the first card, 51 distinct ways of being dealt the next, 50 ways of being dealt the third card, 49 ways of being dealt the fourth, and finally 48 ways of being dealt the fifth. So, there are 52 × 51 × 50 × 49 × 48 distinct ways of being dealt all five cards. But this number is too large. It assumes that the hand was dealt in one particular order, But in what order? It does not matter. You could have been dealt the ace first, second, third, fourth, or last. Fixing when the ace was dealt leaves four possibilities for the king, three for the queen, two for the jack, and one for the ten. So, to compute the number of ways that the hand could be dealt, we must divide (52 × 51 × 50 × 49 × 48) by (5 × 4 × 3 × 2 × 1) to get 2,598,960. It means there are 2,598,959 chances of NOT being dealt the hand A♣ K♣ Q♣ J♣10♣, and one way to be dealt it. But so are the odds of drawing some worthless hand. Anyone would agree that the hand 3♠ 6♥ 8♣ J♦ Q♠ is dull. The odds of being dealt that dull hand are also 2,598,959 to 1. Think about it this way: the odds of *your* being dealt A♣ K♣ Q♣ J♣10♣ are much smaller than the odds of that same hand being dealt to just anyone.

The Birthday Problem

There are at least two mathematical models that give us proper ways to assess coincidences. One is the birthday problem, which tells us that in any group of twenty-three people, the odds are even that two people in the group will have the same birthday. The other is the monkey problem, which asks: if given a large enough amount of time, could a monkey, randomly hitting the keys of computer keyboard, write the first line of a Shakespeare sonnet?

The birthday problem has been famously bandied about the web and in popular math books, and is one of the most explored curiosities in the classroom, so it may seem as if that problem is wildly overdone. Yet it is also the model for thinking about coincidences, and probably the best model we have for doing so. Perhaps we should think of it as the *coincidence problem*; after all, we are asking for the likelihood that two happenings A and B coincide in a large group of space-time happenings. We might ask how large the larger group of happenings has to be for A and B to have a better than even chance of coinciding. The problem is also generalizable enough to give insight into how probability laws work against intuition. The standard problem may be stated this way: In a group of N randomly selected people, how large should N be to give a better than even chance that two people in the group will share the same birthday? The answer is $N = 23$, a surprisingly small number.

Finding N is not difficult. Let $p(N)$ denote the probability of N people *not* sharing the same birthday. First suppose that $N = 2$. Then $p(2) = 365/365 \times 364/365$, because one of the two people can be born on any one of the 365 days, eliminating one day for the other person. This $p(2)$ is very, very close to 1. No surprise there. Next, suppose that $N = 3$. For a reason similar to the $N = 2$ case, the third person cannot share a birthday with any of the other two, so $p(3) = 365/365 \times 364/365 \times 363/365$. This product is easy to compute on a calculator. Continuing in this way, we see that $p(N)$ shrinks as N increases. Eventually we come to $N = 23$, and at that point we are calculating:

$$p(23) = 365/365 \times 364/365 \times 363/365 \times \ldots \times 343/365$$

$$= (1/365)^{23} \times (365 \times 364 \times 363 \times \ldots \times 343)$$

$$= 0.4927.$$

Table 8.1 and Figure 8.1 show that $p(23)$ (the probability that no two of the twenty-three people in the group have the same birthday) is equal to 0.4927. Translating the negative to the positive, we find that the probability that two people in a group of twenty-three have the same birthday is 0.5073, a better than even chance.

Table 8.1.

N	2	3	4	5	6	7	8	9	10	11	12
p	0.9972	0.9918	0.9836	0.9836	0.9595	0.9435	0.9257	0.9054	0.8831	0.8589	0.8330

N	13	14	15	16	17	18	19	20	21	22	23
p	0.8056	0.7769	0.7471	0.7164	0.6850	0.6531	0.6209	0.5886	0.5563	0.5243	0.4927

Even in such a carefully structured problem, there are assumptions that skew the solution. A minor assumption was to ignore leap years. A more major assumption was to ignore the fact that birthdays are not distributed so randomly throughout the year as we think. We know that birthdays tend to cluster for

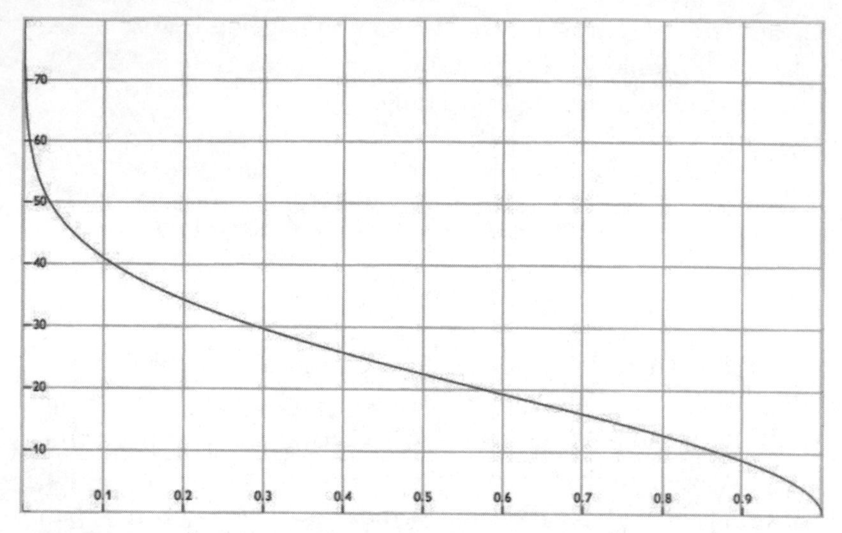

Figure 8.1. Graph relating the size of a group it takes to have no two people share a birthday and the probability that no two people in that group share a birthday.

reasons that might have to do with holidays, natural disasters, seasons, and other unfathomable imbalances.

There are some curiosities. To have a better than even chance that three people in a group share a birthday, you might think that it would take a number close to an additional twenty-three people. The correct number is 88. For four shared birthdays that number becomes 187.[1] Table 8.2 and Figure 8.2 show how the numbers grow, where k represents the number of shared birthdays.[2]

The standard birthday problem was first introduced by Richard von Mises, a Galician-born applied mathematician, who smartly left Berlin in 1933 for a position at the University of Istanbul, where he did great work in fluid mechanics, aerodynamics, and probability theory. In 1939 he came to the United States after accepting an offer at Harvard.[3]

The problem has many disguises. From one point of view, it is a problem of combinatorics. We might even look at it at

Table 8.2.

k	2	3	4	5	6	7	8	9	10	11	12	13
N	23	88	187	313	460	623	798	985	1,181	1,385	1,596	1,813

Thanks to Bruce Levin for this table. ***Source*****: Bruce Levin, "A Representation for Multinominal Cumulative Distribution Functions,"** ***Annals of Statistics*** **9 (1981): 1123–1126.**

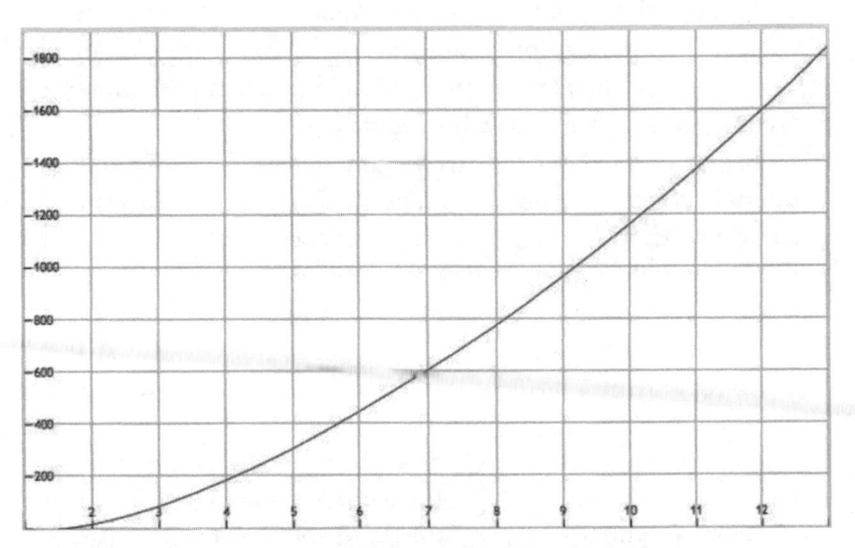

Figure 8.2. Graph relating the size of a group it takes to have a better than even chance that x people share a birthday.

it as a purely hypothetical dice problem: you toss a 365-sided die twenty-three times and ask for the probability that it landed twice on the same side. (It is a hypothetical thought experiment because there is no such thing as a fair physical 365-sided die.) Yet another way of viewing the problem would be to number all the days of the year and scramble them into a random pattern. Perhaps the numbers 1 through 365 could be printed on plastic chips, placed in a rotating cage, and picked *N* times, one at a time, with replacement. Then ask: What is the probability $p(N)$ that one number will be picked twice after N picks?[4]

If we were to change the problem by asking for the number of people who meet at, say a national conference, who share the last four digits of a Social Security number, we would have a similar question. The only difference would be that the number 365 would have to be replaced by 9999, assuming that no person has number 0000. With this assumption there is a better than even chance that in a national conference of 118 participants, two will share the last four digits of their Social Security number.[5] Those last four digits have no real significance and are more or less independent from one's birth date.

Just as I was about to start writing this book, Agnes, a contributing writer to a woman's online magazine, somehow learned that I was writing a book on coincidences. "Dear Professor Mazur: Please forgive what may seem to be a rather odd request," she wrote in an e-mail message to me. "How likely is it to meet someone (really meet, not research online) who shares your birth date (not birthday). It has happened to me twice, and, ironically, at significant moments of my life."

Until that time I had never thought about her complex question. Yet, upon reflection, I quickly gathered that its analysis gives us the essential mathematics of almost all coincidences. Agnes is not asking for the probability that *any* two people in a group share the same birthday; rather, she is asking for the probability that *she*, herself, shares a *birth date* with someone in a group, a much harder question to answer. To distinguish Agnes's question we'll call hers the *birth-mate problem*.

How does one go about answering? We are not talking about 365 days anymore, but thousands of days. What are the variables? Her question is not about the birth dates of any two people, but about *her* own birth date coinciding with someone else's from among her acquaintances. And, making it much harder, it's not just that she has acquaintances that share her birth date; it's about bumping into her birth-mates and discovering that they *are* birth-mates.

If Agnes were interested in calculating the probability that someone she knows shares her birthday, it would be surprisingly easy to give an answer. Let's say her birthday is July 1. Her actual birthday is not important for the problem. It is only a question of specifying a date, or, in other words, wording the problem so that it asks for the probability that someone else in the room has a birthday on a specific date. The chance that one acquaintance was not born on, say, July 1 is 364/365. The probability that N of her acquaintances were not born on July 1 is $(364/365)^N$. So, to calculate an even chance that N of her acquaintances do not share her birthday, we must solve the equation $(364/365)^N = 1/2$ to arrive at N. When we do, we find that $N = 252.65$.[6] And so, Agnes has a better than even chance of sharing her birthday with one of her 253 acquaintances. But that is still a birthday problem, not a birth-mate problem. Agnes's problem goes further. The coincidence for Agnes involves both her birthday and her birth year. For the sake of simplicity, let us assume that most of her typical acquaintances will be within ten years of her age; in other words, ±3,650 days. To have a better than even chance of bumping into one of her birth-mates, she must bump into more than 5,105 acquaintances.[7] That might seem like a lot of meetings. As an active professional woman, she surely encounters 5,105 acquaintances in a five-year span, less than three people a day. But, for the sake of argument, let's make the chances smaller. If we only wish that she should have, say, a 10 percent chance, that number shrinks to 770 meetings. The question then becomes how many of her distinct acquaintances does she encounter over, say a five-year period? Moreover, Agnes has to bump into at least 770 acquaintances *and* have some signal that one of them shares a birth date.

Suppose she runs into $N > 770$ distinct people in a five-year period, and that for some subset of those N chance meetings, the topic of conversation leads to information about birthdays. The difficulty in assessing the complete problem is *not* that there is one person in 770 out there who is her birth-mate, but rather that she found that out inadvertently by having a conversation that led to her knowing that that person is her birth-mate. What are the chances of that? The difficulty in answering is in estimating how often she enters conversations about birthdays. Let's say that on average over a ten-year period that, for every hundred conversations she enters, one conversation turns to birthdays. We must therefore multiply the number of distinct acquaintances by 100. In other words, just to have a 10 percent chance of learning that one acquaintance is a birth-mate, she would have to bump into 77,000 acquaintances. To have a better than even chance of encountering just one birth-mate, she would have to encounter 510,500 of her acquaintances. But Agnes tells us that it happened to her twice! Moreover, the two encounters did not come from her regular acquaintances, but rather from inaugural meetings. The first was with a midwife who delivered her daughter, someone who, as a matter of routine, had to ask for her date of birth. The second happened a dozen years later, when she was in a limo on her way to pick up her parents at Newark Airport. In conversation, she told the driver that her parents were visiting her for her fiftieth birthday. "To further the problem," she later wrote, "those two birth-mates were both professionals that I had never encountered before, and they were not necessarily part of the (intended large) group of my acquaintances likely to be closer to me in age."

So, by any account, we must agree that her two-time encounter was truly amazing.

What applies to birthdays applies to death days. A real case is that three presidents—John Adams, Thomas Jefferson, and James Monroe—died on July 4. Hmm . . . John Adams and Thomas Jefferson died in the same year, 1826. That does seem spooky. However, in their day July 4 was a seriously important milestone day. We know that deaths can be advanced or delayed by hours or days simply by a person's will to live or die. So perhaps those early presidents just hung on to be around for July 4, especially Adams and Jefferson, who held out for the fiftieth anniversary of the signing of the Declaration of Independence. So, there is an element of cause in that randomness. No coincidence.

Monkey Business

The monkey problem began as a statistical mechanics question in the theory of probability, first appearing in *Mécanique Statistique et Irréversibilité*, a 1913 article by Émile Borel. That's the theory that tells us that a monkey randomly hitting keys on a keyboard will type out the complete works of Shakespeare if given enough time. Of course, *enough time* might mean infinite time. English physicist Sir Arthur Eddington was more generous to randomness when he was invited to give a Gifford Lecture at the University of Edinburgh in 1927: "If I let my fingers wander idly over the keys of a typewriter it 'might' happen that my screed made an intelligible sentence. If an army of monkeys were strumming on typewriters they might write all the books in the British Museum."[8]

For now, let's keep things simple. Let's not expect the British Museum, not the complete works of Shakespeare, and not even a sonnet, just the line *shall I compare thee to a summer's day?* If a monkey were to hit the letters *s-h-a-l-l-I-c-o-m-p-a-r-e t-h-e-e-t-o-a-s-u-m-m-e-r-'-s-d-a-y* in that order, we would surely consider it a grand coincidence. What is the likelihood of that? Very slim indeed! A monkey has a 25 to 1 chance of typing the first letter of *shall*, assuming the keyboard is limited to just lowercase English letters. And since one key hit is relatively independent of any others,[9] her probability of typing the first five letters are just 26 × 26 × 26 × 26 × 26 = 11,881,376, or odds of 11,881,375 to 1. But that is the chance of getting it on just the first try. The poor thing should have more than one chance. Much more. Consider

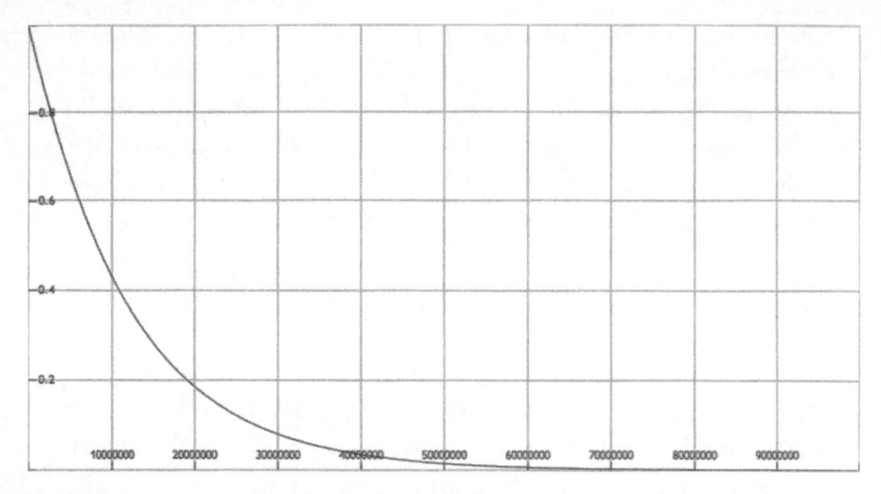

Figure 8.3. Graph of the probability of not typing five specified characters after about *n* tries.

the probability of her not getting it on the first try. That would be, $1-(1/26)^5 \approx 0.99999991583$, close to certainty. After N tries, the probability of her not hitting the keys is $(1-(1/26)^5)^N$.

At $N = 8{,}235{,}542$, she would have a better than even chance of typing the first word in Shakespeare's famous sonnet. Figure 8.3[10] shows how the probability of not typing *shall* comes close to zero after about 50 million tries.

Apply this to password protection. It tells us that a computer program that randomly checks letters could easily break a password with five letter characters. These days, even a relatively slow computer central processing unit could go through 50,000,000 tries in less than ten seconds. But if you put just one more character, the better than even chance of breaking the password would come after 214,124,096 tries. With each additional character (including mixing letters, numbers, and symbols or changing the case), the difficulty grows exponentially. See Figure 8.4.

The probability of randomly keying the first six digits of π on a keypad is 0.000001, or chances of one in a million. There

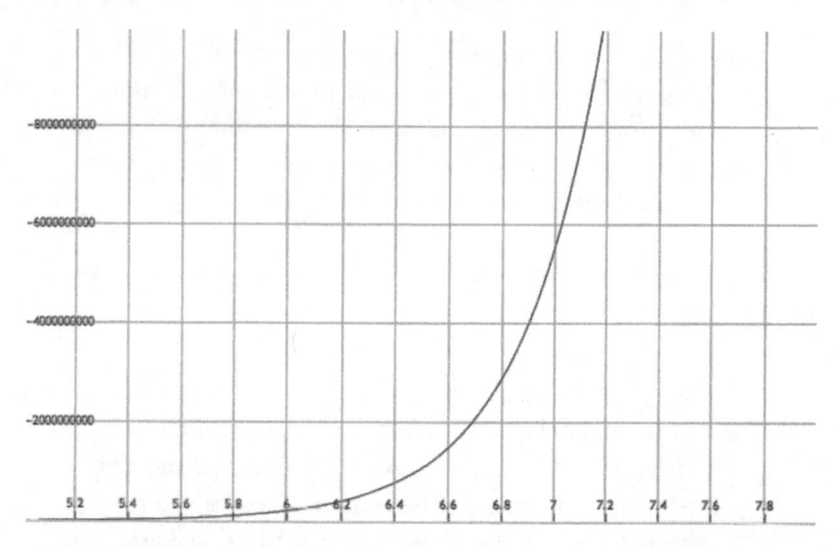

Figure 8.4. Graph showing number of tries to have an even chance of randomly breaking a code with *n* characters.

would be a better than even chance of one in a thousand monkeys keying in the first six digits of π, if each monkey were given a thousand chances. Perhaps π is not such a special number after all. Of course, we are only taking the first six digits of π. Take the first hundred digits of π. With each grain of sand and every star of the universe randomly choosing digits till the end of time, the probability of writing out π to a hundred digits would not be shaken from almost zero. In 1913 Émile Borel asked us to conceive of a million monkeys; randomly striking keys of a typewriter for ten hours a day.[11]

> Les contremaîtres illettrés rassembleraient les feuilles noircies et les relieraient en volumes. Et au bout d'un an, ces volumes se trouveraient renfermer la copie exacte des livres de toute nature et de toutes langues conservés dans les plus riches bibliothèques du monde.

> (The illiterate foremen would gather the blackened sheets and would connect them in volumes. At the end of one year, these volumes would contain exact copies of the books of any nature and all languages preserved in the richest libraries of the world.)

And Sir James Jeans wrote in his book *The Mysterious Universe*:[12]

> It was, I think, Huxley, who said that six monkeys, set to strum unintelligently on typewriters for millions of millions of years, would be bound in time to write all the books in the British Museum. If we examined the last page which a particular monkey had typed, and found that it had chanced, in its blind strumming, to type a Shakespeare sonnet, we should rightly regard the occurrence as a remarkable accident, but if we looked through all the millions of pages the monkeys had turned off in untold millions of years, we might be sure of finding a Shakespeare sonnet somewhere amongst them, the product of the blind play of chance. In the same way, millions of millions of stars wandering blindly through space for millions of millions of years are bound to meet with every sort of accident, and so are bound to produce a certain limited number of planetary systems in time. Yet the number of these must be very small in comparison with the total number of stars in the sky.

Virtual monkeys have simulated the monkey question. On August 4, 2004, computers worked as randomly typing virtual monkeys for 42,162,500,000 billion billion monkey years before typing "VALENTINE. Cease toIdor:eFLP0FRjWK78aXzVOwm)-';8t . . . ,"[13] Astoundingly, the first nineteen characters of this gibberish are precisely the first nineteen characters of the first line of Shakespeare's *The Two Gentlemen of Verona*—

VALENTINE: Cease to persuade, my loving Proteus:

I wondered about the nine consecutive capitals before considering that the caps lock might have been on for a short "coincidental" time. Granted, 42 quintillion is a mega-huge number, but just because it took that long to hit those nineteen characters in that particular order, it does not mean that it couldn't have happened much sooner. Admittedly, it would be an unimaginable quirk to have it happen on the first try, but not impossible. The unexpected could happen, and does happen. Take DNA matching. Are there two unrelated individuals in the world who carry fully matching DNA? The likelihood is unimaginably small, yet not impossible. In fact, the chances are a mere 1 in a billion.

PART 3

The Analysis

Encounters

There are those encounters
that we all seem to share,
opportunities that
we did not know were there
with connections so strong
that tell us who we are,
of why we are here,
and of who's beside us
in those vast cosmoses
of sudden surprises.

—J. M.

THE STORIES IN Part 1 that represent fairly distinct consistent categories are analyzed here:

Story 1: The Anthony Hopkins Story (*Class: Unexpectedly finding what is searched for*)
Story 2: The Anne Parrish Story (*Class: Forgotten objects unexpectedly turning up from the past in faraway places*)
Story 3: The Rocking Chair Story (*Class: Perfect timing and nonhuman chance meeting*)

Story 4: The Golden Scarab Story (*Class: Dream coincidences in somewhat generous time and space*)

Story 5: The Story of Francesco and Manuela (*Class: Unlikely human meetings in precise timings*)

Story 6: The Taxi Driver Story (*Class: Chance meetings of humans under generous timings and spacing*)

Story 7: The Plum Pudding Story (*Class: Repeated encounters and associations with rare objects*)

Story 8: The Windblown Manuscript Story (*Class: Coincidences dictated by natural causes*)

Story 9: Abe Lincoln's Dream (*Class: Prophetic dreams*)

Story 10: Joan Ginther's Lottery Wins (*Class: Spectacularly good or bad gambling*)

Chapter 9

Enormity of the World

We know that the world is large, but we cannot conceive its true enormity. When my daughter Catherine was just eight years old, I would sometimes play a game with her to give her some impression of the vastness of the earth and a sense of number scales. One time she sneezed, and so I asked her to guess at how many people in the world had just sneezed. She would guess at a number as low as two hundred, not a bad guess for an eight-year-old. To her amazement, I would guess at a number in the tens of thousands, probably far lower than the actual number by several digits, given that the size of the human population is now well past 7 billion. Today, a far harder question would be that of barcode readings, those blip sounds that one hears continuously at supermarket checkouts. Take a rough guess at the number of barcode scans that happened during the time it took you to read this sentence. My guess is that you greatly underestimated. The number of scans worldwide exceeds 5 billion a day. It means that, in the time of reading that sentence, nearly one hundred thousand items had been purchased, and that does not include online purchases. Now, that may help us come near some rough sense of the size of the world. But even the number of barcode scans per second is small in comparison with what goes on at a more molecular level.

Nothing is 100 percent certain in this real world of atoms and molecules. Therefore, we must have a way of determining, not what's certain, but rather what is probable. Surely, we might accept without a shadow of doubt that the earth will turn and the sun will rise tomorrow, but most of the world's expected phenomena are accepted by the collective human experience. Theoretical mathematics of an idealized pair of dice could predict the behavior of real dice thrown by a real person. The dice are imperfect white cubes with rounded edges, doubtlessly made in such a way that the indented black dots do not disturb their rotational symmetries. Manufacturers must account for the six slight gouges of the six black dots depleting material that could tilt the cube to bias it toward one.[1] Casino dice are manufactured under highly regulated tolerances. Their expected mean is far closer to 3.5 than ordinary board-game dice.

The law of large numbers is an impressive catch that binds mathematical theory to physical phenomena. It is responsible for many wonders of our fantastic universe as well as Nature's entropic ways of bringing matter and energy disorder to inert uniformity. It even suggests that many of the vast outcomes of the universe are merely results of colossal successions of dice throws and coin flips.

It is easy to believe that events come together in time and space, not by random chance, but by some kind of organized destiny. Is that so? Take the case of how ink disperses in water. One single drop of ink in a bottle of water will uniformly change the color of all the water in the bottle. Is the ink destined to defuse uniformly throughout the bottle, or does the color change uniformly by chance alone? Let's suppose the color is blue. At first you will see a teardrop of blue ink descend from the dropper. If the drop does not splash the water on contact, you will see a blue sphere descend as it morphs through fascinating shapes. It will turn into a torus. That torus will stretch to become a square torus with spheres at its corners. The spheres will split off to become four tori. Those four tori will repeat the process to give 16 tori. The morphosis and splitting will continue until the shapes either hit a wall or hit the bottom to break up. The physics predicts this beautifully by considering all the forces on the spheres and tori. So, the colored ink has a predictable destiny, ordered and organized by physics (i.e., the surface tension of the coloring, the pressure/buoyancy relationship between the two mediums, buoyancy vectors pushing upward, and the velocities of molecules) and the mathematics of the shapes. But when those shapes hit walls, something new takes over. Surface tension is compromised, molecular bonds are jarred, symmetry is broken, and a random element is introduced. At that moment there is turbulence between the two liquids that create a new morphosis, one with an infinitely small likelihood of returning to any symmetry. It is a diffusion of molecules stretching the liquid bonds in seemingly random directions.

What happens if the drop creates a slight splash? In that case, you will see a sphere slowly descend and disperse into magnificent shapes, like cirrus clouds in a gentle wind. Within minutes, depending on the depth, the water will be uniformly blue, a diffusion of ink with no shape at all.[2] Although there is an absurdly small chance that it might return to its original shape, that chance is so infinitesimally small that we can effectively ignore that possibility. Nobody has ever reported seeing it happen. The probability of that improbable coincidence would be a number so small that the number of zeros after its decimal would be larger than the number of grains of sand on earth. But that does not mean it cannot happen. As a model, the phenomenon distinguishes the forward direction of time. The teardrop was in the past and the uniformly blue water is in the present.

What really happened in the bottle to get water from clear to blue? If we look at the question on a molecular level, we understand that each molecule of blue colored ink is not just wandering aimlessly among the water molecules. There are bonds

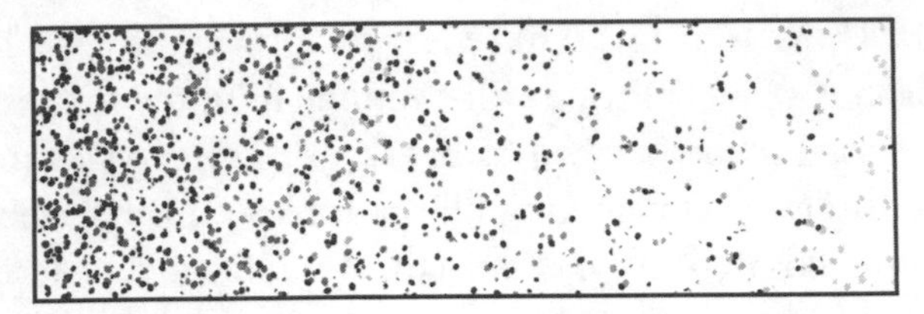

Figure 9.1. Diffusion of particulates in cold water.

that keep the molecules attached, but whatever direction the molecules take, they move in some orderly motion masquerading as random.

What would happen if the molecules had looser bonds? To answer, we change the experiment. Instead of ink, we use very finely ground coffee. Pour a small amount of very finely ground coffee on the left side of a rectangular dish of cold water. Figure 9.1 is a schematic representation of what will happen on the almost microscopic scale. The dots indicate the concentration of coffee grounds diminishing from left to right. Wait a few seconds to see what will happen. The density gradates from left to right, from higher concentration to lower, until it is uniformly distributed throughout the dish.

You might think that there is some force that is driving this tendency of grinds to move from the more crowded region to the less. There is no such force. The grinds have no preference as to where they should go. Every grind in this system is independent of all the others. Each grind is being knocked about by impact with the water molecules and thereby thumped in an entirely unpredictable direction. The path of any grind is determined randomly, or at least as random as anything in real life can be. To understand what is happening, place an imaginary line across the tank dividing the high- and low-density sides, and ask how probable it is that a grind on the imaginary line will move toward the right. The answer is that it is equally likely that it will move to the right as to the left. More grinds will move from left to right than from right to left, merely because there are more grinds on the left side of the imaginary wall than on the right. So, the diffusion toward uniformity occurs simply because there is an equal likelihood that the molecules will move in any direction. It is what happens on a Galton board (see Figure 5.3).

The second law of thermodynamics tells us that we can play the same game with gasses. Take two containers, one with gas at some pressure, the other empty. Connect the two containers by a tube that lets the gas move freely between them. The gas will quickly spread until both containers have half the starting pressure. This equalization of pressure is one example of a universal tendency of particles to distribute themselves in as many ways as possible. Here is the surprise: The gas molecules will randomly bounce off each other like bubbles in a pot of boiling water so that, over time, each will find itself, for a while, back in the container it started from. Henri Poincaré demonstrated this in a general theorem about dynamical systems.

Imagine what would happen if you placed a large number of fleas at the center of a checkerboard. Very quickly the fleas would begin to jump in all directions to fill the checkerboard. Like the fine coffee grounds in that dish of cold water, the fleas are simply jumping around without any predetermined direction. Any one flea is not jumping to have more space for itself; for even if it has lots of space, it will jump again in a new random direction. Fleas spread themselves out by their random jumps. Will they ever return to their original places, if they continue to jump? Perhaps not; however, consider the following thought experiment: Imagine two containers. One, labeled *A*, contains one hundred balls, each labeled uniquely with the numbers 1 to 100. The other, labeled *B*, contains nothing. Also imagine a bucket of chips uniquely numbered 1 to 100. Randomly pick a chip and read its number, *N*. Take the ball numbered *N* from bucket *A* and place it in bucket *B*. Replace the chip and repeat the process. Each time the chip *N* is picked, move the ball labeled

N from whatever bucket it is in to the other bucket. Can you guess what will happen? Yes, the number of balls in container *A* will decrease exponentially until both containers have approximately the same number of balls. But as the number of balls in container *A* decreases, so does the likelihood of picking a chip with a number from container *A*. In fact, the rate of decrease is proportional to the number of balls remaining in container *A*. Now, I repeat the question: can you guess what will happen in the long run? It may seem counterintuitive, even amazing, but with absolute certainty, all the balls will eventually return to container *A*, though it may take an enormously long time for that to happen. Poincaré's general theorem on dynamical systems predicts this.[3] It suggests, as both Plato and Bernoulli alluded to, *apocatastasis*, "that after the unrolling of innumerable centuries everything would return to its original state."[4] The late Sir James Jeans, a renowned physicist who was knighted for his contributions to astronomy and popularization of physics, used to quip that anyone still breathing today is breathing in the molecules of the dying breaths of Julius Caesar.

These examples work because we are dealing with a large number of objects. When numbers are exceedingly large, like molecules in a drop of ink, or human populations spread over the vastness of this planet, we have a better chance of averaging the random element, and of knowing what might happen to an individual in the crowd.

Very many complex phenomena of nature may be simply explained as flipping a coin or randomly picking a number a gazillion times. And from that huge volume of random numbers, chance creates an ever-evolving dynamic world, a world where colored ink diffuses in water with no ultimate purpose, where gas shares pressure with a vacuum to satisfy the laws of thermodynamics, where fleas jump aimlessly yet spread out on checkerboards, and where DNA erroneously replicates itself without a plan, and so, fortuitously, creates distinctive humans.

Hidden Variables

Hidden variables deceive us into thinking that causes are either not there, or that they are too hard to find. The enormity of the world plays a part, along with all the invisible strings that link its parts. We think in local terms without regard to the manifold interactions between the parts that make up our world, from subatomic particles to galaxies.

Sometimes two totally independent variables seem to have some statistical connection through a third variable. When that happens, we see an illusory correlation from the way we see the data or the way the data is presented. If we were to naively collect grades and hair length of students in a math class, it is likely that we would have a direct correlation between hair length and grades. Those with long hair are likely to get good grades. If we didn't look at the third variable, we might conclude from that correlation that students should grow their hair to get a good grade in that class. We are not so naive to miss a third variable, say, age or gender. Hair length could have been skewed toward the older students who just happened to have longer hair or toward the women who had longer hair than the men.[5] Another example would be a correlation between income later in life and college grades. We might confuse it to wrongly conclude that income later in life depends on an individual's school grades, when in fact the hidden variable is the amount of hard work and time a student was willing to experience.[6]

Hidden variables are ubiquitous in statistical data correlation. Without spotting those variables we are bound to mistakenly believe all sorts of nonsense such as to get good college grades one should start smoking because "cigarette smokers make higher college grades than nonsmokers." Or take a more serious example: Until quite recently, in the New Hebrides of the South Pacific, it was believed that body lice led to good health. For centuries the elders casually noticed that healthy natives had lice and sick natives often did not. They concluded that lice caused good health. Under a more careful and controlled study, it was observed that almost everyone had lice most of the time. Lice could also cause a fever that would result in their self-destruction because of that fever. The confusion was over the fact that unhealthy people were the ones who got fevers and were lice-free. "There you have cause and effect altogether confusingly distorted, reversed, and intermingled," Darrell Huff wrote in his now more than sixty-year-old and still best-selling book *How to Lie with Statistics*.[7] The media is filled with all sorts of strange takes on what we should believe from polling studies: pesticides in farmlands cause autism; power lines cause brain tumors; wasabi-root tea is a muscle relaxant; 9 out of 10 doctors agree that cereal for breakfast promotes good health; children with longer arms reason better than do those with shorter arms; and walking in a pine forest once a week decreases the stress hormone cortisol, blood pressure, and heart rate. Women should take estrogen to diminish their chance of heart attacks. Estrogen therapy increases the likelihood of heart attacks for women who already had heart disease. Estrogen therapy might protect women against osteoporosis and perhaps colorectal cancer, but it might also increase risks of heart disease, stroke, blood clots, breast cancer, and dementia.[8]

There is the classic case of Sir Ronald Aylmer Fisher's blunder. For many bioscientists and statisticians, Fisher is the father of modern statistics and experimental design. He was born in 1890, in a London suburb, and died of colon cancer in 1962, in Adelaide, Australia. Richard Dawkins called Fisher the greatest biologist since Darwin.

Fisher was a man of great charm and warmth, a probing thinker of wide interests, a man of passionate devotion to scientific investigation, a stimulating conversationalist, but also one that occasionally displayed an irrepressible temper toward whomever he found guilty of making, causing, or disseminating errors. His writings are obscure, and so was his teaching: "Fisher was too difficult for the average student; his classes would rapidly fall away until only two or three students who could stand the pace remained as fascinated disciples."[9]

Early in Fisher's career as a statistician, he worked at an experimental agricultural station, a place that would later become world renowned for the development of experimental design. He developed what is today called the analysis of variance, established a principle of randomization, and advanced the importance of replication.[10] He designed experiments to test a coincidence by quantitative techniques that involved matching the cards of an ordinary deck of fifty-two playing cards, to systematically study extrasensory perception.[11] It is a practical method that calls for a scoring system based on permutations of the deck that distribute themselves normally.

It's hard to believe that a biology genius like Fisher could have encouraged work in eugenics, the misguided notion popular before the 1930s that, unless governments encouraged the birthrates of families with "desired" genetic traits and discouraged those with "inferior" traits, genetic stock would contribute to the decline of civilization.

In August 1958, Fisher wrote in the journal *Nature* that "The curious associations with lung cancer found in relation to smoking habits do not, in the minds of some of us, lend themselves easily to the simple conclusion that the products of combustion reaching the surface of the bronchus induce, though after a long interval, the development of a cancer. If, for example, it were possible to infer that smoking cigarettes is a cause of this disease, it would equally be possible to infer on exactly similar grounds that inhaling cigarette smoke was a practice of considerable prophylactic value in preventing the disease, for the practice of inhaling is rarer among patients with cancer of the lung than with others."[12] Fisher viewed arguments linking lung cancer with smoking as unconfirmed suppositions.[13]

> The subject is complicated, and I mentioned at an early stage that the logical distinction was between A causing B, B causing A, something else causing both. Is it possible, then, that lung cancer—that is to say, the pre-cancerous condition which must exist and is known to exist for years in those who are going to show overt lung cancer—is one of the causes of smoking cigarettes? I don't think it can be excluded. I don't think we know enough to say that it is such a cause.[14]

Fisher's work was flawed. Given his temperament of irrepressible temper toward anyone he thought made a mistake in analysis of data or judgment, one can only imagine how infuriated he would have been with someone who had made the mistake he had, in prematurely drawing conclusions and not examining all the data available. He was not recognizing his own personal and professional conflicts: he was a smoker hired by the tobacco industry.

Unfortunately, the results of too many health studies generate speculations about causes and prevention that end up too quickly in the popular media. We are given recommendations to eat more fish and less trans fats, and to not live close to electromagnetic fields. Such public health recommendations could lead to other dangers. We were once told that to diminish our chances of heart disease we should take vitamins C and E, and beta-carotene as antioxidants. To prevent colon cancer, we should be eating more fiber. We were once told to have a low-roughage diet, and then a few decades later, to eat plenty of roughage. Other large-scale observational studies could not confirm these theories. Just because a clinical study involving tens of thousands of subjects in trial and in control trial confirms a hypothesis, it does not mean that one event causes another. All it can do is provide a possibility that a hypothesis is correct. At best it only gives circumstantial evidence that one event causes another. Without knowing the cause for sure, we know very little about how to make specific recommendations. Indeed, if the cause is wrong, the recommendations might do more harm than good.[15]

It's not as if these clinical studies do not tell us anything. They tell us plenty. For example, we definitely know that cigarette smoking has some causal association with lung cancer and cardiovascular diseases, even though we do not know the actual cause. Cigarette smoking *is* one of the contributory causes. We know this from the coincidental spike in women's cancer rate during WWII, when American women suddenly entered the workforce by the millions and started smoking for the first time. We have some indication that American diet and lifestyle has some connection to breast cancer from examining Japanese women and American women and then two generations of Japanese American women who end up with the same breast cancer rates as American women. The problem is that causation is not a simple idea. There are often confounding circumstances that make us think that one thing is a cause of another: when A causes B indirectly because A actually causes C, which happens to cause B.

The problem with clinical trials is that they are not as random as they should be. Nobody ever asked me to be the subject in a clinical trial. So, we must wonder: who are those subjects? They are people who are motivated to volunteer. Many are paid, and many are paid by sources that might have some connection to the interests of the funders. Therefore, the subjects come from a very special group, not a randomized group. The kinds of people involved in clinical trials are those who would be more faithful to sticking to recommendations that will be self-beneficial. They are likely to be thinner, and have fewer health risks. We can statistically adjust for the effect of socioeconomic status, but that does not always work so well.[16] Moreover, the findings of these studies are temporary, waiting for another decade or two before the next study comes along to question the earlier one. In other words, clinical study biases are very hard to avoid.

On the other hand, if the public listens to health advice coming from clinical studies, we learn something. If we were wrong in accusing cigarette smoking to be the cause of lung cancer and cardiovascular disease, we should not have seen the dramatic decrease in lung cancer and cardiovascular disease that we did see over the last five decades, during which the smoking population in the United States dropped by 57 percent.

History tells us that what we believe now may not be believable a century from now. There is more out there than just the things we see, the things we measure, and the things we think we know. Our scientific beliefs are confidences of the moment. Samuel Arbesman tells us in his book *The Half-Life of Facts*, "We accumulate scientific knowledge like clockwork, with the result that facts are overturned at regular intervals in our quest to better understand the world."[17] Beliefs, no matter how strong they may be today, are not the last words. They are simply working hypotheses. There is a pinch of randomness in the original recipe of the universe, and our tools for observation are limited; so, we cannot know everything.

Yes, we are limited. Events in nature depend on so many variables that exact measurement is often impossible; and that's ignoring the stickler of the uncertainty principle. If a simple event, such as the flip of a coin, depends on countless undetectable happenings in a mildly chaotic world of accidentally colliding electrons, just imagine the myriads of happenings responsible for such a complex phenomenon as cancer. But discovering the cause of cancer is not the same as having a pretty good guess of what is a suspect. Some scientists attributed the increase in lung cancer in industrialized countries following WWII to occupational factors and new industrial products. Asphalt was suspect, because of the proliferation of road building in America and Europe. However, by the end of 1950, with so many studies linking smoking and lung cancer, it became clear that smoking was a huge factor. The job of statistics is not to find causes, but rather to find suspects. Many natural relations that cannot be explained by laws or measured by observations can be linked by statistical measurements.

Back in the fifth century BC, Hippocrates wrote about a powder extract from tree bark that eases headaches and relieves fevers. It was aspirin. The German pharmaceutical company Bayer has been producing it in tablet form since the nineteenth century. But nobody knew why it worked until 1971, when British pharmacologist John Robert Vane showed that aspirin suppressed the production of certain molecular compounds that regulate the contraction and relaxation of muscle tissue. Morphine has been around since the sixteenth century as a painkiller, but before 2003 no one knew that it occurs naturally in the human body. We should think about all those good practices we do before knowing why we do them. Long before anyone knew about bacteria, people washed hands before eating. These days, it is possible that we wash too often, even with antibacterial soap, which kills the beneficial bacteria. How were we to know that some bacteria benefit our health?

Science loves to know the direct links between causes and effects, but it does not require us to know that there are such links. Scientists may suspect a correlation between two complex phenomena. The real problem is that humans naturally tend to make connections where there are none, and also tend to ignore connections that are too complex to predict. We see coincidences as events that are mysteriously fated by some deeply significant design. That might be true, and it might not be. In a highly complex world of interconnected phenomena, some connections are so subtly coupled through long chains of indirect links that we can never envision the effect of one on another.

Chapter 10

The Stories of Chapter 2 Revisited

COINCIDENCES ARE DISTINGUISHED stories that arouse our attention to probability. No one doubts that such stories are exceedingly rare, but how rare does a story have to be to shrink the world in time and space? The following stories are indeed rare, though inevitably likely to happen.

Story 1: The Anthony Hopkins Story

The Hopkins story might simply be one of synchronicity. Just think of how many places *The Girl from Petrovka* could have been. Think of the number of other people who could have picked up that book before Hopkins saw it. Think of why Hopkins found a book by that title and, furthermore, that very copy—the one belonging to George Feifer. And then consider the possibility of Hopkins sitting right next to it and not noticing it: a close version of the story—perhaps a better version—would have happened just the same, but Hopkins would never have known about it, and we would never have heard about it. One reason the story is so compelling is that it involves a particular person, moreover, a celebrity figure. It is by every measure a spectacular story, mostly because we know the person to whom it happened. Is the Hopkins story really such a spectacular coincidence? We have some sense that it is, but from where does that sense come? It might be spectacular, but what information do we have to back it up? There are no numbers to give us an impression of likelihood.

Yes, it might be synchronicity. But to clarify the difference between synchronicity and mathematical plausibility let's look at some numbers: the number of books that are left behind in train stations, the number of bookstores there are in central London, and the number of people that come to town every day in search of a particular book. The story happened in 1976. That matters, because back then there was no Internet, no Amazon, to make roaming for books so easy. Back then, the easiest thing you could do was to phone each bookstore to save yourself a whole lot of time of going to them physically.

To analyze the Hopkins story, we must take into account the vastness of the city of London. At the time of this writing, and in the age of the Internet, there are 111 small independent bookstores in London. To survive, each of those stores must attract on average at least ten buyers per day. By a conservative estimate, those stores must collectively sell at least one thousand books a day. A more realistic estimate would put the number at about three thousand. Some people come to browse, others come to look for a particular book that they intend to buy, and still others just come to get out of the rain or to pass some free time. Let's say that just one hundred people come each day to buy a specific book of the title X.

It's not likely that any one of those one hundred people will find the book they are looking for sitting on a bench in an Underground station. But let's take the opportunity now to think about how many people accidentally leave books in public places and how many people simply abandon the books they've finished reading on trains and in railway stations before their trains depart.

If book X has any reasonable popularity when it's first released, it will sell at least one thousand copies in its first month.

What happens to those copies? Some will end up unread and left on a bookshelf in a person's home. Others will be sold to used bookstores, and still others will be left in public places.

My guess is that *The Girl from Petrovka* had sales above ten thousand. That would give the law of large numbers headway in showing that the Hopkins event had between a slim and a reasonable chance of happening, at least happening to someone. How is that? Suppose ten books were left behind in public spaces in London, some on park benches, others in cafés, waiting rooms, hotel lobbies, and so on—a very reasonable estimate. Let N be the number of people coming to London looking for one of those books. Those N people are more likely than not to be noticing books left on public benches. So, the question becomes: what is the probability p that such a person will see the book he or she is looking for? How do we get p? Unfortunately, unlike tossing a die or dealing with a deck of cards, this scenario is not easily amenable to computing that p. Knowing p exactly is almost impossible.

There is a way, though. We could create a computer model that simulates the wanderings of people near and far from what they are looking for. It would not be an easy task, because of the many hidden variables that connect real people's thoughts to their experiences. But such a model would give us a numerical approximation to the mathematical probability p, a number that is—for now—hidden from our comprehension. A simpler way is to create a mental picture that relies on our intuitive sense of how people behave when wandering city streets while searching for something. Yes, that plays into the dangers of biased subjective feelings, but it does make us think about the problem more deeply.

We leave the actual story, the one involving Anthony Hopkins and George Feifer, and get some sense of how likely anyone coming to central London to search for a book finds it left behind somewhere in a public place. This is a much easier task. If we find that likelihood, and it turns out to be very small, then we know that the actual story involving Hopkins and Feifer is far less likely. So, we will be doing what mathematicians often do: put upper boundaries on the numbers we want to find, in this case, boundaries to the probability that a book searcher will successfully find the book he or she is looking for. We will also do something else that mathematicians often do: simplify the problem to clarify matters, recognizing that the real problem, to be dealt with later, is far more complicated.

London is a large city with 60,000 streets, more than 3,000 small parks and garden squares, 8 large royal parks, 111 bookstores, and 276 Underground stations scattered throughout the city. However, if we return for a few moments to the Hopkins story, we can limit the area to more workable numbers. Hopkins said that he found the book at an Underground stop near Hyde Park Square. Feifer confirmed that he gave the book to a friend who lost it near Hyde Park Square. The nearest Underground station to Hyde Park Square is Marble Arch, which is almost a straight-line half-hour walk along Wigmore Street to the vicinity of the British Museum, which at the time of Hopkins' story was the largest bookshop neighborhood of London. It makes sense to limit the search and wanderings to, say, a 2-mile radius centering on the British Museum. In that area there are nearly one thousand streets. But many are very short with very few bookshops, and few book searchers would wander off the main roads. Moreover, abandoned books are more likely to be in more traveled places, such as Underground stations, and leisure spots, such as parks.

The core of the story is not about Anthony Hopkins, and not about *The Girl from Petrovka*. It is about someone's finding a particular book on a particular day in an extremely unexpected place.

So, let's imagine N people walking to and from bookstores in hopeless search of the books they came for. Let's limit their meanderings to a 2-mile radius of the British Museum. Further, let us suppose ten books have been left in public spaces within the area. Will any of those N people accidentally find the particular book they came for among the ten abandoned books? Probably not, if N is a small number. This is a very rough thought-experiment model, but not so rough as you might think, because book searchers are not taking random paths through London. They are more likely to spot any abandoned book in an unusual place. Now, let N be some large number. We expect that before the end of a day's walks, $k \leq 10$ abandoned books will be spotted, and therefore we should have an approximate success ratio k/N. In other words, there would be k successes in N tries. The weak law of large numbers then says that this success ratio is a pretty good approximation to p when N is large enough. So, the question becomes: what N is large enough? Surely $N = 10,000$ would give us a pretty good chance that k would be greater than zero. Nobody would expect ten thousand people would randomly roam the streets of London one day, searching for books, even though Greater London has a population of over 8.6 million. However, if we extend the time limitation to one year, and assume that one hundred people are searching each day, many as repeat searchers, then $N = 36,500$. For two years $N = 73,000$. With that more liberal acceptance of N, there should be close to a better than even chance that some one of those 73,000 will find the book he or she is looking for. Of course, why just two years? Why not ten? And why just London? We could take the whole United States, with its 22,500 bookshops, or the entire world. This wonderful law of large numbers teaches us to not underestimate the size of the world.

This is a creative model, but not one that tells the full story. Hidden variables are everywhere. Even people in search of particular books might easily be in the vicinity of their hunted book without ever seeing it. Moreover, we can see that N would have to be enormous, far more than 73,000, for any one of those N

people to be matched up with the very book they are looking for. So, yes, the likelihood of that happening is surely much smaller than whatever k/N we imagine.

But the weak law of large numbers tells us that the difference between p and k/N will be as small as we wish, if N is large enough. We could instinctively guess that if $N = 73{,}000$ (two years' worth of searchers), then k would be at least 1, and then bravely assume that N is large enough to make the assumption that $P[|k/N - p| < 0.001] > 0.5$. That would tell us that there is a better than even chance that the likelihood of one person's finding the book he or she is looking for would be close to 0.000014, which gives odds of 71,427 to 1, very close to the odds of getting a straight flush in poker!

All this means is that the upper bound of the real probability is not terribly low. The likelihood of the real story, of its happening to a named particular person, is far smaller. So, though we are left with no definite numerical likelihood that the original story is fantastically rare, we do have some sense that stories like it are not so rare.

The big question is not that Hopkins found a copy of *The Girl from Petrovka*, but rather that it happened to be Feifer's copy! Now, that is a real coincidence, with a p incomprehensibly small. Except! . . . Except that Feifer said that he lost his copy close to the area in which it was found.

Story 2: The Anne Parrish Story

The Anne Parrish story is different. Parrish was just browsing, not searching for any book, let alone her own. After analyzing the Hopkins story, we can see that the Parrish story is less rare.

If we know nothing about Anne Parrish's life, her story seems astounding. It is a fabulous story with no apparent cause. Alexander Woollcott, a literary critic for the *New Yorker* at that time, who knew Anne Parrish, wrote the story when Ms. Parrish was still alive. Here's Woollcott—

> When we thus catch life in the very act of rhyming, our inordinate pleasure is a measure, perhaps, of how frightened we really are by the mystery of its uncharted seas. At least, I know when I first heard the tale, I carried it about with me as a talisman, more than half-disposed to believe that when the obvious Anne Parrish crossed the street to that book-stall, somewhere in fathomless space a star chuckled—chuckled and skipped in its course.[1]

But let's connect the dots. Her mother, whose name was also Anne, but called Année, studied painting at the Pennsylvania Academy of Fine Arts in 1860 at the same time as Mary Cassatt. While at the Pennsylvania Academy, Année and Mary became close friends. Mary became a famous impressionist portrait painter and went on to live, study, and work in Paris, befriending Edgar Degas and Camille Pissarro. Now, could it be that Année passed the book on to her good friend Mary, who took it to Paris? Mary died in 1926. Her estate was probably dispersed, along with her library, and Anne Parrish's American book "probably" ended up in on the tables of the Paris bookstalls sometime between 1926 and 1929, before Anne Parrish found it.

So, let's think further about this. If you were an American visiting Paris in 1929, chances are that sometime on your visit you would come to both Shakespeare and Company and the bookstalls on the Seine. Those were the known places that buy and sell used, nonrare English books. If you were primarily a children's book author, you would likely be rummaging the shelves of children's books. In fact, most authors I know of rummage bookstore shelves—in particular shelves of the genre they write—whenever they get a chance. So, here we have a very likely chain of links connecting *Jack Frost and Other Stories* on the tables of Seine bookstalls all the way back to a young girl for whom *Jack Frost and Other Stories* was a favorite book.

But hold on. As with all good coincidences, timing was essential. Anne had to have been in Paris at a time when the book was at that Seine bookstall. Had she come earlier, or after someone bought the book, she would have missed the opportunity. Perhaps another American would have bought it, brought it back to America to give Anne another chance. But that would have been a different, less astounding coincidence quite possibly weakened in a forever-hidden history of the book's journey to Paris and back. Timing here had a wide margin to give likelihood a good advantage.

Assigning numerical odds would be difficult. But let's take some reasonable guesses. First, let's guess the likelihood that Anne would be traveling to Paris in the summer of 1929. I would give that likelihood a conservative number close to 0.1. Anne was married to an industrialist with money. Paris was the number one European vacation destination of wealthy Americans in 1929, along with sailing tours of the Greek Islands. What is the likelihood that she would visit the bookstalls while in Paris? I would say the likelihood of that is 0.3. The hardest to peg is the likelihood that the book would be there. Now here is where the background story helps—Anne's mother's connection to Mary Cassatt, Mary's death, and the few places in Paris that would have dealt with used English books. I would guess that that likelihood would be near 0.01. So the probability of such a story happening would be something like $p = 0.1 \times 0.3 \times 0.01 = 0.0003$, the odds in favor of it happening are 3,331 to 1. Unlikely, but not as low as the odds of coming to a city with the purpose of finding a particular book and finding it on a public bench. Yes, there are plenty of unaccounted hidden variables that complicate our estimate, but they would not change the probability by more than 1/10,000, and therefore the odds of the Anne Parish story remain slightly better than the odds of being dealt a poker hand of four of a kind.

Story 3: The Rocking Chair Story

Anne Parrish's story had the advantage of loose timing. *Jack Frost and Other Stories* could have been among the English books at the bookstall for months before and could have remained there for months ahead, had Anne chosen another time to come to Paris.

The rocking chair story is a type that could happen only under precise timing. The details of the story, as presented in Chapter 2, are as follows: A rocking chair sat in my brother's living room in Cambridge, Massachusetts. My wife ordered one of the same make and design from a store in Cambridge. The chair was out of stock and therefore was to be delivered to my brother's house at a later date, several weeks later. At a small gathering at my brother's house, a guest sat on his rocking chair. Just seconds after it collapsed and broke into many pieces, the doorbell rang. The new chair was delivered.

As with any of these stories, the numerical odds are hard to know. But we can get a handle on at least the level of odds.

The story could very well be a case of synchronicity. But consider the variables: The rocking chair ordered was an exact duplicate. That fact contributed to the story, but not to the coincidence. My wife had seen the rocking chair in my brother's living room and wanted that exact chair. She was probably told where to buy it. The first contributing variable was that the chair was not in stock. Had it been, there would be no remarkable story.

The second variable was the visit by the guest. His being in my brother's living room at that moment might have been pretty probable. He was a friend who visited often, so we might safely estimate the odds of his being there are better than 9 to 1, and therefore a probability of p_1, where $0.1 < p_1 \leq 1$. There are, of course, the odds that he will choose to sit in the rocking chair. That's an easy one to figure. As far as I recall, there were two couches that could accommodate six people, and four chairs, including the one black rocking chair. If choosing a seat were random, and if no one else had been sitting yet, the odds of his choosing the rocking chair would be p_2, where $0.1 < p_2 \leq 0.01$. But people don't choose their seats in a room randomly, especially when a rocking chair is one possibility. So, without knowing anything about the person, those odds are hard to establish. For argument's sake, though, let's agree that $0.1 < p_2 \leq 0.01$.

It's hard to estimate the timing of the break—that is, the probability that the chair would break at the moment the guest took his seat. All we can do is assume that the chair was about to break. It's a liberty that we take with the understanding that in the end we have to give our estimate some liberal play.

The timing of the delivery is somewhat easier to peg. If the chair was out of stock and the delivery was expected in the next two weeks, we should expect it to arrive sometime in the second week during business hours. There are 3,360 business minutes in a week. We could take things down to the second the doorbell rang, as the story says, but to avoid the question of any possible exaggeration of the details, let's keep things to minutes. The humor of the situation stands just as well. So, the probability p_3 that the doorbell should ring in that particular minute when the guest sits and the chair breaks is 1/3,360, or approximately 0.0003. We may therefore conclude that the probability $p = p_1 \times p_2 \times p_3$ of the story's happening to that particular group of people is between 0.0000003 and 0.0003. Surprising any intuition, this story is staggeringly unlikely. The odds are between 3,333,332 to 1, and 3,332 to 1. At the least, it is worse than winning a lottery with one of four tickets. At the most, it is better than being dealt four of a kind in poker.

Story 4: The Golden Scarab Story

Scarab (or Scarabaeidae) is the name given to a family consisting of a particular classification of beetle. Their large body, metallic colors, and club-shaped antennae distinguish them. June bugs and Japanese beetles are just a few of the most common seen in the United States. Carl Jung once had a patient tell him of her dream about a golden scarab. Sitting in a chair with his back to a closed window and listening to the dream, he heard a gentle tapping on the window. He turned to see a flying insect tapping the windowpane from the outside, as if to get his attention. He opened the window and caught the insect as it flew in. It was indeed a scarab. Jung took this coincidence to be a model example of what he called synchronicity, the simultaneity of two events that come together in time and space by means that cannot be explained by chance.

If the dream of the golden scarab is an example of synchronicity, then we would not be able to know the odds of its happening. It falls into a category different than the rocking chair story, but like that story, it is also one of critical timing. Had the scarab tapped on the window a half-hour later, the story would have been different. There might very well be a synchronicity in the universe, but this story surely involves chance. That being said, we should keep in mind that the young woman's dream brings in the hidden variable of the collective unconscious, which cannot be ignored.

June bugs are common in June. A June bug could have tapped on the young woman's window as she was having her dream. Had she heard it in her sleep it could have affected her dream. Our dreams are often a mix of unconscious and conscious experiences sometimes influenced by real sounds and lights. A person could sleep through a real thunderstorm while dreaming about being in the middle of one. So, the question for us is this: what are the chances that a scarab tapped on her window during her dream? And what are the chances that a scarab tapped on Jung's window at the same moment that the woman was telling her dream?

Jung does not tell us what time of year the encounter happened. It could have been June. Judging from my own scarab encounters, I would say that the answer to the first question is roughly 29 to 1. I encounter at least one June bug tapping on my window screen at least once a year, and almost always in June. The answer to the second question is more challenging. The odds of a scarab tapping on Jung's window is also 29 to 1, but that does not take into account the very important precise timing of two other events, the interval in which the young woman is having her dream and Jung's encounter with the scarab at his window. And that's the puzzle for which we must make assumptions. Attracted to light, June bugs tap on windows mostly at night. The fact that her dream was significant enough to tell about it at her session with Jung gives evidence that it was a rare dream that was possibly interrupted by a window-tapping scarab. If we take the conservative position that she could have dreamed that particular dream on any of the June nights, the probability of it happening on the same night as the scarab visit would be $1/30 \times 1/30 \approx 0.001$, or odds of 998 to 1.

Let's suppose that the patient had an appointment with Jung once a week for one hour. And let's suppose that Jung would have seen on average six patients a day, excluding weekend days. That's 132 one-hour visits in the month of June. The scarab dream was told on just one of these visits, and told in, say, a ten-minute segment. There are 792 such segments in June. That means that over the month of June, the likelihood of a scarab coming to the window at the time of the telling of the dream would be 791 to 1, giving a probability of 1/792. The probability of the story happening is therefore $1/30 \times 1/30 \times 1/792 \approx 0.0000014$, less likely than a royal flush!

Story 5: The Story of Francesco and Manuela

The Francesco-Manuela coincidence is not the story itself, but rather the fact that a person writing a book about coincidences was there to hear it from the person who owns that story. Look at it this way: Those particular names, Francesco and Manuela, do not matter. The story could have been about any names, say, Bill and Joan, or Fred and Fredrika. The story could have happened at any place in the world. It didn't even have to be about two men and two women. Conceptualize the story and we find that, in the abstract, it is about two pairs of people, each with the same pair of names, meeting anywhere in the world for the first time. Now the story becomes one of counting pairings of names. How many names are there in this world, and how many of those pairs will meet at some time in, say, a year? We can't even begin to guess those numbers. In Olbia alone, a city of 58,000 inhabitants, there were, at the time of this writing, 2,834 persons with the name Francesco, and 276 Manuelas. But one thing is certain: the number of pairs of people in the world with pairing names is great—in fact, enormous! Such a mistaken identity story as the one just told cannot be so unusual. What is more unusual is that we have two pairs of people who spend an extravagant amount of time unaware that they are at the wrong meeting. Granted, such obliviousness significantly brings down the numbers. The limitations we just imposed bring those numbers down to at least hundreds.

There are some loose methods that might lead us to good guesses about the odds. With 2,834 Francescos in Olbia, we must wonder how many Manuelas are visiting Olbia from Madrid on any given day. How many of them stay at Hotel de Plam, where the story starts?[2] And how many are in the lobby of Hotel de Plam to meet someone they had never seen before? We could measure the likelihood that tomorrow morning two people by the name of Manuela will be waiting in that hotel lobby to meet two people named Francesco, people they have never seen before. We could do it by spending mornings in the lobby, asking people their names and canvassing them about whether they are there to meet someone not seen before. Then, over a ten-day period we could take the daily average number of people named Manuela who are sitting in the lobby, and divide that by the daily average number of people just sitting in the lobby. That number might be zero. But if we increase the number of days to 365, that number of people will more likely turn out to be more than zero. Of course, this is a time-consuming, expensive way to measure a probability.

There is another way. Start with the average number of people visiting Olbia on any given day. Sardinia is an island, so visitors must come either by sea or air. Take air. Before September 2013, there was one nonstop flight on Iberia Airlines. But just after my wife and I left the island, Olbia was flooded by a storm that left half the city in ruins. The nonstop flight was canceled and never renewed. By finding the number of one-stop flights from Madrid (10) and the average number of passengers on Airbus 320s and 340s making those flights (200), we learn that on average two thousand people visit Olbia every day from Madrid. And since Olbia is pretty much a destination site, almost everyone who comes does not board another plane that day. There are, of course, fluctuations from summer to winter. From a sample of the Madrid telephone directory, we find that 1.3 percent of the Madrid population are named Manuela. We then make the rash but conservative assumption that only a quarter of the passengers on those ten planes coming from Madrid (500) were residents of Madrid and its environs. From that we learn that on any given day Olbia hosts 6.5 new visitors named Manuela. It is possible that some then take a train or bus to a different town. So, let us conservatively guess that we are down to three new visitors. Now there are many arguments to be made about where those visitors would stay and what sort of person would choose what sort of hotel. My analysis limits the average number of persons by the name of Manuela staying at Hotel de Plam down to 0.17. As long as we are talking about averages, we might as well suggest that the hotel choices are clustered—some hotels offer special deals on certain days and certain times of year. One Manuela could have arrived in Olbia the night before. Another may have just arrived. Considering those clusters and arrival times, the odds of two Manuelas choosing the hotel suggested by their respective Francesco hosts is 35 to 1, just the same odds as rolling boxcars (double sixes) with a pair of dice. Should it be surprising to find two Manuelas from Madrid in Hotel de Plam? I leave it to you to answer that. The real issue for the coincidence is how it happened that the linkages between the Francesco/Manuela pairs got mixed up for such a long time before any one of the four people involved became suspicious that something was wrong. For that I have no answer, other than to say that people who do not know each other normally have awkward introductory conversations that do not at first center around the actual purpose of their meeting.

Was it a remarkable coincidence? The mistaken identity event is more common than we think because numbers behind them are larger than we imagine. Our analysis considered just two names, Francesco and Manuela. The story surprises us, not because of those particular names, but rather because I heard the story from Francesco himself.

Frame the story differently: Someone by the name X is to meet someone by the name Y in the lobby of hotel H. Another person by the name X is to meet another person by the name Y in the lobby of H. So far, it is just a variation on the famous birthday problem, which we came across in Chapter 8. However, it goes further. Each person is misidentified for one hour. Now the possibilities are far greater. Just examine what happens if X and Y stand for one of four different names, say, X = Marco,

Andrea, Francesco, or Luca (the four most frequent male names in Italy). Likewise let Y = Maria, Laura, Marta, or Paula (the four most frequent female names in Spain). And, of course, for the meeting of interest, neither X nor Y has to stick to the names of any particular gender. Now we see that their chances of such a meeting are very much increased. We now have sixteen possibilities: the Marcos could meet the Marias, or the Lauras, or the Martas or the Paulas. And that goes also for Andrea, Francesco, and Luca. In the end, we have sixteen more chances of a mistaken identity meeting in the lobby of hotel H.[3] Why not take the first hundred most popular names in Italy and the first hundred most popular names in Spain? If we let n be the number of name pairs, we might speculate that the effect grows as the square of n. That would mean that with one hundred name pairs, the chances multiply by ten thousand. However, as the popularity of names diminishes on the list of popular names, so do the numbers of people with those names. If we limit this analysis to, say, $n \leq 25$, it is safe to say that the effect grows roughly as the square of n. That's a factor of 625. But hold on. There are close to 51,733 hotels of three stars and higher in Italy. And if we include all the hotel lobbies of the entire world, our number would grow so large that we should be certain that two pairs of people will have a mistaken identity meeting in some hotel lobby (I would guess) somewhere every hour!

"Now wait a second," you say, as my wife does, too. "Francesco told the story to *you*. There is a difference between the likelihood of a mistaken identity event like the Francesco-Manuela meeting and an arbitrary meeting of two unidentifiable people somewhere, someplace in the world. The coincidence is not only that it happened, but that you were told about it." Yes, I agree. However, by the analysis above, it should be happening someplace in the world multiple times a day. Isn't it surprising that I heard the story just once in my entire life? Why should I be surprised to hear it at all if it is so inevitable?

Every one of the coincidence stories in this book can be analyzed by looking at numbers. The difficulty lies in finding the many significant hidden variables. The numbers might not look big at first, as they didn't in the case of the Francesco-Manuela meeting, but by careful examination of all the possible interacting combinations of events, those seemingly small numbers grow to be quite large—large enough to make something that seems impossible into something that is inevitable.

Story 6: The Taxi Driver Story

A woman hails a taxi in Chicago. Three years later she hails a taxi in Miami and finds that her driver is the same one she had in Chicago. To explain this, we should first examine the frequency of her calls for taxis. The woman is an executive of a private equity firm, someone who frequently takes taxis in different major cities. Taxi drivers that do not have albinism are not as distinguishable; so, a person who uses taxis often might expect to hail a taxi without noticing that the driver is familiar, unless that driver happened to be a person with albinism. So, it's possible that she had twice hailed a different driver in two different cities without being aware of doing so.

Let's consider the probability of her hailing a taxi in Chicago and Miami three years apart with the same driver—named A—without regard to whether or not the person had albinism. The probability of hailing A in Chicago is 1, because taxis are not driverless, yet. We first estimate the probability that a taxi driver in Chicago moves to Miami within three years. Today there are 15,327 taxicab drivers in Chicago, and about 5,000 in Miami. Statistics on how many people leave Chicago for Miami are not available, so all we can do is look at the exodus numbers. We have data that shows 95,000 people out of a Chicago population of 2,722,389 moved to other states in 2014. That's a proportion of 1 in 29 per year. If that same proportion holds for the 15,327 taxicab drivers in Chicago, then we may daringly presume that 529 drivers moved to other states within a three-year period. Chicago is the third-largest city in the country and Miami is the forty-fourth. It's tough to guess their destination cities; however, the U-Haul list of top destination cities in the United States ranks Miami number 40. So, we could assume that very few Chicago taxicab drivers moved to Miami, perhaps more than twenty and less than forty. That puts the woman's chances of hailing A greater than 20/15,327 = 0.013 and less than 40/15,327 = 0.026. The odds are between 75 to 1 and 36 to 1. Not bad!

Now back to the driver with albinism. Since we did not take into consideration whether the woman would notice that two taxicab drivers on two occasions three years apart, the odds must be the same. The trick, as with all coincidences, is in the noticing.

Story 7: The Plum Pudding Story

The plum pudding story, as told by nineteenth-century French poet Émile Deschamps, cannot be reduced to any justifiable numbers. It ranks as one of the greatest coincidence stories I've ever heard, partly because of the great length of time between the connecting incidences. On the one hand, that length of time increases the chances, and on the other it enriches the story. The key circumstances are these: Young Deschamps first met Monsieur de Fortgibu while tasting plum pudding, a dish that was almost unheard of in France at the time.

Ten years later, after having forgotten about the plum pudding, Deschamps passes a restaurant that displays plum pudding on its menu. He enters the restaurant to order a portion, but is told by the counter waitress that there is none left because a man in a colonel's uniform had ordered the whole pudding. She points to M. de Fortgibu. Years pass once again, in which Deschamps does not see or think about plum pudding. Then one day he was invited to a dinner at a friend's. Plum pudding is served, and

Deschamps tells the hostess and guests the story of M. de Fortgibu and the plum pudding, as if it were a fantastic coincidence. Just when Deschamps completes his story, the doorbell rings and M. de Fortgibu is announced. The same M. de Fortgibu, invited to a different dinner at a neighbor's, had mistaken the address and rung the wrong doorbell.

This story falls into a category close to chance meetings, but we are talking about four variables coming together in space and time in such a way that there is such confounding that untangling the variables would be almost impossible without wild presumptions. The number of years that pass between events, makes the problem almost impenetrable. Almost, but let's try to break this story down to numbers. The probability of meeting M. de Fortgibu over a dish like plum pudding the first time is 1. The particular person and the plum pudding have no real relevance. The story could have centered on a different person and a different noun. Finding the odds of the second meeting ten years later is more challenging. Deschamps could have walked passed the restaurant without noticing that plum pudding was on the menu. But that would not have been likely, for plum pudding to him was something special, not like noticing *mousse au chocolat* on the menu. So, it is very, very likely that he would take notice, and slightly less likely that he would enter to order a portion. The coincidence is that M. de Fortgibu was there.

Let's look at it this way: Paris in Deschamps's time, the mid-nineteenth century, was a small city; not in population, but more in where people would be. Certain quarters of Paris were more frequented by certain people than by others. Had M. de Fortgibu passed the restaurant, he, too, would have noticed the sign and very likely entered to order a portion of plum pudding. The behavior is much like noticing a taxicab driver with albinism. You notice something more when it is uncommon, and when it stirs your recollection of things past. The other thing to keep in mind is that it is entirely possible that M. de Fortgibu dined at that restaurant every day, just as it is also possible that it was Deschamps's first time dining there. So, as far as this first coincidence is concerned, it was a chance meeting of two people with a common interest in a relatively small geographic area. It is the next coincidence that brings us to something hugely uncommon and immensely difficult to analyze: M. de Fortgibu mistakenly ringing the doorbell of the apartment where Deschamps is dining and where plum pudding is being served.

However, this coincidence happened many years after the restaurant meeting. We have to consider all the years that M. de Fortgibu did not mistakenly ring the doorbell of someone hosting a dinner with Deschamps as one of the guests, whether or not plum pudding was one of the foods being served.

Story 8: The Windblown Manuscript Story

Late-nineteenth-century French astronomer and popular science writer Nicolas Camille Flammarion told this story. He was writing an eight-hundred-page popular treatise on the atmosphere. While writing a chapter on the wind force, a sudden gale blew open a window, lifted the whole chapter from his desk, and carried it out to a soaking downpour of rain. A secondary coincidence happened several days later, when a porter for his publisher, who worked 1 mile from Flammarion's apartment, fortuitously found the missing pages of the chapter and brought them to him.

It might seem astonishing that the wind could have blown all the papers so far to coincidentally wander from 32, avenue de l'Observatoire to the offices of Flammarion's own publisher, Librairie Hachette, at 79, boulevard Saint-Germain. But there is more to the story to give us some causal background. The slightly hidden part of the story is that on the morning of the wind event, the same porter came to Flammarion's apartment to deliver some page proofs.[4] The man lived near Flammarion, and went to breakfast just after delivering the page proofs. On his way back to the offices of Flammarion's publisher, he spotted the sodden leaves on the ground and—noticing the handwriting to be Flammarion's—thought that he had accidently dropped them. He returned to his office and told nobody for a few days, presumably letting the leaves dry. So, in this case, the cause was that the person who found the pages already had a close connection to the person who lost them.

The gale that came while Flammarion was writing about wind force is not so surprising. Anyone writing a chapter of a book does not do so in a matter of minutes. He could have been writing for days or weeks. Open windows on summer days are notorious at sucking papers into any winds and breezes that come along. So, the main event is the coinciding blown pages and porter. The porter lived in the neighborhood, was familiar with Flammarion's handwriting, was in the literary business (and therefore would have been interested in what the papers were about), and was an occasional visitor to Flammarion's apartment. These suggest somewhat favorable odds that the papers would be found and returned. But those odds are dampened by the higher likelihood that someone else would find the papers, someone who did not know Flammarion's handwriting, or a street sweeper who would have put them in trash containers with the other street refuse.

Story 9: Abe Lincoln's Dream

Lincoln told of his dream of hearing a group of sobbing mourners, and leaving his bedroom to find out where the sobs were coming from. The mourners were invisible and the sounds were all around. When he came to the East Room, he saw a corpse in a catafalque surrounded by several military guards and weeping mourners. He was told that the president was assassinated.

He had many premonition dreams. When the war began, he had the same dream before every important national event. Were they coincidences, or simply understandable anxieties of uncertainty surfacing in subconscious and surfacing in a dream state?

Lincoln's dream of his own assassination could simply be recognition of uncertainty of his position. No US president had been assassinated before, but that doesn't mean that assassination was not on his mind, especially in time of war. Like most dreams, premonition is built into the dream mechanism; we are still "thinking" while we are dreaming, or we are "thinking" that we are dreaming.

Story 10: Joan Ginther's Lottery Wins

Joan Ginther won four lotteries. She won $5.4 million the first time, $2 million the second time, $3 million the third time, and $10 million the fourth. Her wins were spread over an eighteen-year period, starting in 1993. I admit that the chance of it happening to her is hopeless, but not impossible. Technically, her story is not a coincidence. Coincidences have no apparent causes. Ginther's story has a definite cause: she picked winning numbers by buying tickets in bulk. We might think that her four-time lottery win was a colossal bit of luck. Of course, it was. Those multiple wins are indeed rare. But there are hidden factors.

First, her first win gave her house money to play again and again, each time using her gambling losses to cover part of her tax debt to the government. It was smart, but that's what 80 percent of jackpot winners do: play again and again hoping for that next rush. Gaming theory psychologists refer to those rushes as reinforcement of favorable history.[5] And when you are a jackpot winner, you don't just buy a ticket or two; you buy them by the hundreds, and even thousands. But how does someone pick winning numbers?

It was reported that the odds of picking those four lottery winning numbers are 18 septillion to 1, and that it is so unlikely that it should happen to that person only once in a quadrillion

Table 10.1. Texas Lottery Odds

Match	Amount	Average odds	Probability	Expected value
6 numbers*	Jackpot*	25,827,165:1	0.000000038	$0.09
5 numbers	$2,000	89,678:1	0.00001115	$0.02
4 numbers	$50	1,526:1	0.000654878	$0.03
3 numbers	$3	75:1	0.013157894	$0.02

*** Depends on the number of tickets sold and how many weeks have gone by without jackpot winners.**

years.[6] (See Chapter 7 to see how such a calculation is done.) That may be, but without knowing how many times Ginther lost (and we have no way of knowing), there is no way of knowing the true odds. Some parts of her story are missing. It's true that she has a PhD in mathematics from Stanford, so perhaps she used some algorithm for determining winning numbers while buying in bulk.

Let's consider the Lotto Texas lottery. Players purchase a single ticket for $1 and mark six numbers from 1 to 54. The Texas Lotto lottery advertises the odds as displayed in Table 10.1. Let's suppose Ginther bought a single ticket for $1 and had chosen the winning six numbers. With a jackpot prize of $2 million, the expectation of winning that jackpot is a mere 9 cents on a dollar. It is possible to win one of the other three nonjackpot prizes, so we must add an expected value of 7 cents (the total excluding the jackpot) to the jackpot's expected value, thereby making the expected value for winning any prize 16 cents. For each dollar played, the player is throwing away 84 cents.

Then there are taxes and the possibility of sharing the prize to have the expected value shrink to approximately 12 cents. The pool of players increases with the size of the jackpot, so the likelihood that a winner will share the jackpot increases.

Yes, winning four times with different jackpots is a colossal bit of luck. The probability of one win is spectacularly low. Four wins for Ginther would be so low that its probability would have thirty-two zeros after the decimal point before any digits greater than zero begin to appear. But that is only because we are specifying Joan Ginther as the person who is to win four times. Surely, she has as much chance as anyone else in winning it any number of times, even just once, as long as she's buying just one ticket at a time. But the chances of *somebody* winning the jackpot are quite high, given that close to a billion Lotto Texas tickets are sold each year. After all, someone does win, although it may take a few drawings before a winner comes forward. In 2014 an estimated 31,818,182 distinct people spent over $70 billion on lottery tickets in America. If 70 billion tickets are bought in a year, and the numbers are picked randomly (they are not absolutely random, as we have noted in Chapter 6), then someone is sure to win within a year, and the odds are still pretty good that someone will win within a month.

We can understand how one person can win, but what about the same person winning four times? The odds are quite good that wins like Ginther's have a good chance of happening in a population of almost 320 million Americans. Her wins seem striking only because we are viewing them as happening to one specific person, Joan Ginther.

Let's calculate the probability that a person, *some* person, not necessarily Ginther, will win lotteries twice within a five-year period. You may find the outcome quite surprising. There are 26 distinct major legal lotteries in North America with 104 drawings per year, with a total of 13,520 drawings in a five-year period. On average, 1/6 of the number of drawings result in a jackpot win, so the number of wins is 2,253. We assume that 80 percent of winners continue to buy lottery tickets at each drawing as they did before, at least for five years. Also, on average, the number of jackpot winners per jackpot win is 1.7.

Now we make the rash assumption that these winning events are independent of each other. It's rash because we are assuming that winners of each winning draw continue to bet large sums and use the same strategy as before to affect the next winning. We also make the assumption, just to make the analysis possible, that each winner uses the same strategy as any of the others. In other words, we average the strategies among the jackpot winners. Otherwise the problem becomes far too difficult to analyze.

Let x be the probability that a person who continuously plays the lottery for five years, will win twice. We take p to be the probability of winning the jackpot on a single drawing from Table 10.1. We first compute $(1-x)$, the probability that first time winners will *not* win a second time within five years. Let $y = 1-x$. The number of jackpot winners per jackpot win is on average 1.7, so on each winning draw the number of new jackpot winners increases by a factor of 1.7. It means that on the first of the 2,253 wins, there will be 1.7 winners. On the second of the 2,253 wins, there will be 1.7×2 winners, . . . and at the last of the 2,253 wins, there will be $1.7 \times 2{,}253$ winners. Put another way, the probability that a first time winner will not win a second time at the first, second, third, . . . and last of the 2,253 wins is $(1-p)^{1.7}$, $(1-p)^{1.7\times2}$, $(1-p)^{1.7\times3}$, . . . $(1-p)^{1.7\times2,253}$, respectively. Since we are assuming that each win is independent of any of the others, y, the probability that none of the first-time winners will win a second time is the product $(1-p)^{1.7}(1-p)^{1.7\times2}(1-p)^{1.7\times3} \ldots (1-p)^{1.7\times2,253}$.

Therefore, $y = (1-p)^{1.7(1+2+3+\ldots 2,253)} = (1-p)^{4,316,523} \approx 0.49$. So, $x = 0.51$, or better than even odds that someone will win a jackpot twice in a five-year period.

We can make a similar calculation for the world for a one-year period. There are 166 lotteries in the world. Many non-US lotteries have just one drawing per week; so the number of drawings per week worldwide, including the US biweekly drawings, in two years is 9,984. The number of jackpot wins in a single year (using the scale where, in the United States, the number of drawings to jackpots on average is 5 to 1, and the ratio of drawings to jackpots in the rest of the world is 3 to 1) is therefore 2,496. Using the same method we calculate $y = (1-p)^{1.7\times 2,496} = (1-p)^{5,297,635} \approx 0.40$. Therefore $x = 0.60$.

In two years the probability that a person will win twice is 0.97, a number so close to 1 that the chance of someone winning the jackpot twice in two years is almost certain.

Joan Ginther's wins were over an eighteen-year period. In that time span, the probability of some person winning four jackpots somewhere in the world is extremely close to 1.

PART 4

The Head-Scratchers

Phrases Like These

Some stories defying all rules
can have us believe they are flukes
to further the chance collisions
of feats we can never predict,
those hard to reckon surprises
that taunt mathematical law
so rambling fingers of monkeys,
a million competing each day
frantically typing away
in vast numbers of trials,
a gazillion denials,
with no aim or purpose
of crafting an opus
can crank out a sentence like this.

—J. M.

There are coincidences that completely escape analysis. No matter how you look at them, they seem to come to us through serendipity. They do not fit into any of the ten categories listed in Part 3. The first of these five essays investigates coincidences of DNA evidence at crime scenes and jurors' misperceptions of the remote chances of DNA mistakes. The second presents the story of Wilhelm Conrad Röntgen's accidental discovery of X-rays while experimenting with electric currents in a glass container under partial vacuum. The third tells the story of a rogue trader, Jérôme Kerviel, who gambled with 10 million euros without advance knowledge of two flukes: one that made millions of euros and one that lost many more. Essay 4 is about the psychic powers of extrasensory perception and the question of whether they fall into any coincidence category. Essay 5 compares the planned coincidences of literature and folktales with the unpredictable coincidences of real life.

Chapter 11

Evidence

> It is better and more satisfactory to acquit a thousand guilty persons than to put a single innocent one to death.[1]
>
> —MAIMONIDES

PEOPLE LOVE COINCIDENCE stories and think they are very rare. When many of those same people become jurors in cases that might lead to an execution, they think coincidences of forensic mishap are not easily possible. Still, jurors want strong forensic evidence before they are willing to convict. That's a good thing. Curiously, on the flip side, they are all too often ready to convict in the face of strong forensic evidence of innocence. The general public mistakenly presumes that DNA evidence is the absolute proof of guilt or innocence, at least if it is not compromised by contamination. However, coincidences of criminal evidence that lead to wrongful convictions are far more probable than we might expect.

Arguments of DNA evidence are powerful, especially to folks who have only a cursory understanding of how DNA evidence works. Folks who have little knowledge of DNA's intricacies are marks for courtroom sharks that can shrewdly manipulate beliefs in their favor, as DNA can be contradictorily used both as evidence for conviction and as evidence of innocence in messy investigations of serious crimes. The question of what constitutes DNA evidence—what it can prove and what it cannot prove—is far too complex to give a neat answer. Nevertheless, we must raise the question of evidence to focus on where coincidence is inferred as evidence of guilt or innocence. Mistakes in evidence—circumstantial, coincidental, and material—can contaminate judgments of guilt.

Before DNA testing, blood typing, serology, and conventional fingerprinting were the standard tools. These conventional forensic tools give very imprecise measurements compared with DNA fingerprinting. About 40 percent of Americans share type O-positive blood, and matching fingerprints are inconclusive in many criminal cases. Barry Scheck, cofounder of the Innocence Project and one of the attorneys on O. J. Simpson's defense team, said that DNA identification is "the gold standard of innocence and the magical black box that suddenly produces the truth."[2] DNA fingerprinting is now playing a major role in exonerating wrongly convicted prisoners. Still, defense or prosecution lawyers could play DNA testing to their advantage, either by impressing the jury with its unassailable scientific accuracy or by attacking the evidence collection and storage procedures. In the O. J. Simpson case, the prosecution had substantial DNA evidence; but the defense was able to persuade the jury that the evidence had been tampered with.

DNA fingerprinting is not infallible. There can be unintentional errors and there can be deliberate manipulation. Machinery imperfections, environmental accidents, and human handling failures—all might contribute to erroneous results in the testing lab.

On May 11, 2006, an independent investigator reviewed hundreds of criminal cases that were originally analyzed by the

Houston Police Department Crime Lab and Property Rooms. In seven forensic science disciplines, including serology, DNA, and trace evidence, major mishandling issues were uncovered in cases dating back to 1980. In reviewing 135 DNA analyses, forty-three (32%) were identified as having major mistreatment issues, with a suspicion of intentional scientific fraud.[3]

Matching a DNA profile with samples found at a crime scene is not dependable evidence of guilt or innocence. Take the case of Yara Gambirasio, one well-known example among many. In November 2010 thirteen-year-old Yara went missing from her home in Brembate di Sopra, a small village in northern Italy. Her body was found three months later in another village 6 miles from her home. The investigation went down several dead ends for two years before a match was found. It wasn't a perfect match, but it was strikingly similar to male DNA found on Yara's underwear. The match belonged to a man who was in South America at the time of the crime, but it led to another search in another town and eventually to two postage stamps licked by a man who had died in 1999. "It was just a crazy coincidence," the chief investigator told reporters at one time when she was about to abandon her only promising lead. "There was no connection," she said. "You couldn't make it up. This whole case is crazy."[4] The complete story takes many turns and in the end the crime is solved. The person who happened to be in South America was lucky to have had such a foolproof alibi. Lucky for the dead man to be dead.

Jury members should understand, or at least be instructed by judges to understand, that DNA analysis is an extremely complex and tricky process that could easily result in false positive or false negative identifications. It is inevitable that some piece of information will be interpreted and processed as relevant and positively incriminating when it is simply circumstantial. Any hidden precision to what happened could get lost in the accuracy of how the analysis is interpreted. Similarly, there is always the possibility that some piece of information will be interpreted as exonerating when, in fact, it is truly incriminating.

On one level, DNA analysis requires some uncontaminated biological material from the crime scene—blood, sperm, skin cells, hair roots, saliva, or perspiration. DNA from the environment—plants, insects, bacteria, or other humans—often contaminates the samples. Another issue is our understanding of the uniqueness of the DNA fingerprint. There are questions to ask: How unique is a DNA fingerprint? Is it possible for two people (who are not identical twins) to coincidentally share the same DNA profile? Is DNA analysis perfect? Could there be a false positive or false negative? Even in its purest form there is still a chance—albeit an extremely small chance—that DNA readings from two distinct (nontwin) people are identical. Do we want to take that chance of executing an innocent person when that person is accused and convicted exclusively on the basis of DNA evidence?

As for false positives, which depend on the individual circumstances, their overall odds have been estimated to be between 100 to 1 and 1,000 to 1.[5] There are errors in sample handling. Miscalculating the odds of false positives can lead to the incrimination of innocent people, especially when identified through a DNA dragnet. Labs rarely, though occasionally, misinterpret test results. They may make incorrect reports of test results because there is a possibility of a coincidental match by virtue of a positive random match probability. Regrettably, juries are rarely given statistics on the frequency of false positives. And yet both the odds of a coincidental match (where two people actually do have the same DNA profile) and the odds of a false positive match should both be considered for fair assessment of DNA evidence.[6]

Junk science sometimes gets involved. So many people believe that hair-sample evidence is DNA evidence. It is not. The DNA evidence can be established only with a sample of hair root.

In most forensic cases, hair-sample evidence is based on subjective microscopic observations and comparisons, truly bogus evidence. There is no reliable scientific way of determining the ownership of a hair-sample that does not include the root.[7] And yet, for decades the courts have relied on so-called hair-sample experts for criminal prosecutorial testimony.

Consider the cases of three black men, Donald Gates, Kirk Odom, and Santae Tribble. Their convictions were based on microscopic hair comparison evidence, until DNA analysis contradicted that evidence. In 1990, a jury that heard the prosecution exaggerate the statistical likelihood of a hair sample match convicted Tribble of murder. He was sentenced to twenty years to life. He served twenty-three years in prison before being exonerated, all because of a hair found in a ski mask.[8] A match? What match? Science has yet to come up with any meaningful statistical frequency distribution of hair characteristics in a sample population.[9] So, where is this scientific evidence coming from? How can a recognized expert claim a match when, in the absence of nuclear DNA, there is no scientific way to determine ownership of hair specimens in the larger population? Yet we often hear experts telling the jury that evidentiary hair could be associated with a specific individual: "In my opinion, based on my experience in the laboratory and having done 16,000 hair examinations, my opinion is that those hairs came from the deceased."[10] Anyone can have an opinion. But opinions of experts in the courtroom are often taken as proof. This is not just sheer nonsense; it is irresponsible, given the weight of possibly incarcerating an innocent human being. No one can give a positive statistical probability by microscopic analysis that a particular hair specimen originated from a particular source. And yet, over the last two decades, twenty-six of twenty-eight FBI lab experts emphasized in testimony a near certainty in matching hair samples. In Mr. Tribble's case, one expert claimed a match "in all microscopic characteristics." In its closing remarks, the prosecution hammered home a fabricated and misleading statistic: that there was only "one chance in ten million" that the hair could have belonged to someone other than Mr. Tribble.[11]

Unfortunately, real crimes are not like those we see on TV or in the movies, where forensic analyses always seem to be infallible. More unfortunately, real juries generally believe what judges tell them, and what they hear and don't hear. They hear prosecutors telling them—as one did without objection from the judge—that "the beauty of DNA testing is that it can give you a hundred percent certainty."[12] No forensic test is 100 percent certain, yet people have an ongoing misconception that DNA has a definite yes or no answer. In fact, DNA analysis depends on the validity of the test, and the heritage group connected to the suspect. But the courts accept forensic evidence as if it is rock-solid science, without any full consideration of its limitations.[13] In one Houston Police Department Crime Lab case, the forensic analyst misleadingly testified, "No other two persons will have the same DNA except in the case of—identical twins."[14] Anyone with an educated understanding of how a DNA profile works in the crime lab should know that such a statement is far from true. In all fairness for due process, the jury should have been told that there is always some small percentage of the population that is expected to match the profile. The small probability of a match does not eliminate coincidences. In most cases involving DNA evidence, the jury is typically given statistical data on coincidental matches. The jury is generally told the odds of an unrelated individual selected at random matching the defendant's DNA profile. But those numbers are meaningless to a juror who believes those odds of, say, 1 in half a million means absolute certainty.

The Human Genome

Let us briefly recall a few points about the human genome, the genetic information encoded in the chromosome pairs of each

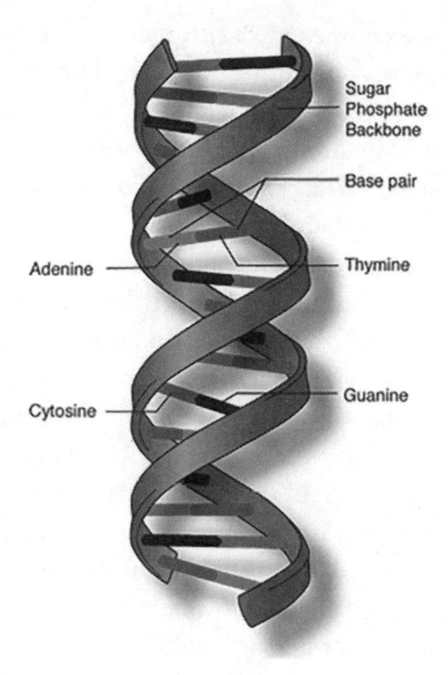

Figure 11.1. Structure of the double helix.
Courtesy of the National Human Genome Research Institute, the National Institutes of Health, and the illustrator Darryl Leja.

human cell's nucleus. A chromosome is a packaging of DNA molecules in the nucleus of a cell. A person has twenty-three pairs of chromosomes (22 pairs, plus two sex chromosomes), paired as one chromosome from the mother's set, the other from the father's. As long as we realize that the full story of genetic information is far more complex than the text of the next few pages, we can get along with a reasonable picture of how to identify a person by his or her DNA.

DNA is the acronym for the chemical deoxyribonucleic acid, found in living cells. Think of the structure of DNA as a spiral staircase, a double-helix staircase (Figure 11.1).

The steps are composed of nitrogen-based chemicals called *nucleotides* or *bases*—adenine, guanine, thymine, and cytosine, more simply denoted by the letters A, G, T, and C. Two spiraling bands composed of paired sugar and phosphate molecules form the sides of the staircase. Each step is a joining of nucleotides from each of the two bands. The arrangement of the letters defines a person's genotype, or genetic identity.

To depict DNA sequences, we first look at short tandem repeats (STRs), which are repeats of a combination of the four nucleotides A, T, G, and C. There are $4 \times 4 \times 4 \times 4 = 256$ possible order combinations. Think of arranging any four sequences of the letters A, T, G, and C, allowing any letters to repeat. So, you would then have AAAA, or AGTC, or any of the other 254 combinations. One person might have a chromosome with an STR that might be AGTT, AGTT, AGTT. Another person might have a chromosome with an STR that might be AGTT, AGTT, AGTT, AGTT. And still another person could have six repeats, or twelve. Notice that the first person had just three repeats, whereas the second had four. This creates far more variation in the genetic print of individuality in humans. And if we throw in the fact that a person inherits a sequence on each chromosome from his or her mother and another from his or her father, the likelihood that two people in the worldwide population (excluding identical twins) have the same DNA is close to zero, but not zero. Just to give some impression of how small and how long a double-helix DNA molecule in a single cell is, consider this: it is packed with the nucleus of a cell that is smaller than fifty thousandths of a centimeter in diameter, and when it is unraveled end to end, it is 2 meters long. That is unimaginably tight packing.

For a sense of the complexity of the model, think about this: in each of the twenty-three pairs of chromosomes are about 3 billion sequencings of the four nucleotides, each from the mother and father.[15] No doubt that is a large number. The problem is that we do not know *which* of the 3 billion sequence positions can vary.

To distinguish the DNA identities of two people with a 100 percent match, we would have to compare about 3 billion nucleotide pairs, an impractical and very expensive process. We do not do that. Instead, we compare a very small portion to find a comparison. If there is a match in that small proportion, we estimate how likely that match could have occurred by coincidence. The question we are left with is this: how small should the "small proportion" be to give us a comfortable feeling that the match did not occur by coincidence?

Forensic scientists collectively agreed on a *random-match probability* based on just thirteen different STRs. That is, they claimed they could identify a person by thirteen different STRs distributed through the human genome. The hope is that nonmatches will show up on this random sampling among the thirteen STRs on the twenty-three human chromosomes. Why just thirteen? That is a decision made on the bases of practicality and expense. Their reasoning is that the number of STRs on each of the thirteen sites would vary enormously among individuals in any human population. For example, on chromosome 3, one person might have inherited five repeats from her mother, and another person might have inherited three repeats from her mother and six from her father. In the larger population, some repeats will be very rare, but some others will be quite common. It takes just one difference to rule out someone as having the same DNA as that recovered from a crime scene. In a single chromosome, the STRs might not be so infrequent. A reasonably low frequency in the population might be, say, 0.1. But multiply that by the frequencies of STRs in the thirteen chosen chromosomes and you find that the probability of a match is on the order of 1 in a million billion. But the list of suspects of a crime is a far smaller group than the population of the whole world.

So, forensic scientists are very confident that—for all practical purposes—there is no chance at all that two people will have the same set of copies. The chance of two people having the same pairs over all thirteen STRs is not zero, but narrowed to a group of suspects for a crime, the chances are so phenomenally close to zero, that we can assume that they actually are zero.

In other words, if the DNA profiles from the crime scene and the suspect are a match, then the evidence points to the suspect's guilt. On the other hand, if the profiles do not match, then the evidence points to the suspect's innocence. That is DNA fingerprinting, and forensic evidence. Whichever way the evidence points, the investigation must also consider that natural flukes, coincidences, human behavior, and unaccountable hidden variables routinely complicate the easy pictures, especially those that come from a single measurement.

The Central Park Jogger

Every guilty verdict of an innocent person is a scar of justice, but the rape case of the Central Park jogger Patricia Meili, with its coincidental timing and crossing paths of a large group of Latino and black teenagers, is a deep mutilation of justice. There was no DNA match, yet five of those teenagers were convicted by confessions of being at the crime scene. They spent six to thirteen years in prison before the real rapist confessed. A prosecutor can use DNA evidence to get a conviction, but when DNA evidence is against a conviction or used for the purpose of exoneration, that same prosecutor may argue, the way some do, that "DNA evidence in and of itself is not always the 'silver bullet' that it is sometimes perceived to be."[16]

The prosecutor's office told its story. On April 19, 1989, a large gang of males stormed through Central Park looking for trouble when they came upon a young jogger. They were called a "wolf pack" and it was said that they had been out for a night of "wilding." It was said that the wolf pack beat Patricia Meili into a coma, dragged her to a ravine, sexually assaulted her, and left her to die. The story was explosive in the press because the accused were all black, and the jogger was a twenty-eight-year old white investment associate "on the fast track" in the Corporate Finance Department of Salomon Brothers. Patricia, or Trisha as she now calls herself, suffered a traumatic brain injury that left her with no memory of the attack. It made a sensational and inflaming story to sell newspapers and to attract and hook TV news viewers, a good racial-tension story. "Mention the Central Park Jogger to virtually any adult in New York City," Trisha wrote in her memoir, "and to millions across the country, and they'll relive their sense of shock at what happened to her, even fourteen years later."

Trisha's jogging route occasionally varied. Sometimes it would take her to the dimly lit areas north of 84th Street. Friends had warned her not to jog alone at night, so she would take the northern route at the beginning of her jogs when the evening was still early. This time she entered Central Park on 84th Street, and turned north to the 102nd Street cross-drive before being brutally attacked and raped. Without any memory, there were no eyewitness identifiers, no evidence of who could have done it, nothing but identification of the proximity of the whereabouts of people at a particular time.

The tale is gory with no need to go into the particulars. For a while Trisha was fighting for her life; then, once in stable condition, it seemed as if she would suffer permanent brain damage from the gruesome way she was beaten. She had suffered an extreme brain swelling that was forecast by physicians of Surgical Intensive Care of Metropolitan Hospital in East Harlem to result in "intellectual, physical and emotional incapacity."[17] No one completely recovers from a rape, especially from a brutal one. But Trisha did physically recover. Her life took a different direction from investment banking

The beating and rape was pinned on a group of five black and Hispanic teenagers. Detectives and prosecutors coerced them into signing documents containing incriminating evidence that was admissible in court. They were just boys who knew nothing about their rights as public citizens. Their path just happened to be near Trisha at the time of the rape. For that, they were convicted in 1990, even though DNA samples taken from Trisha's panties did not match any of the samples taken from the accused.

In 2002 Manhattan district attorney Robert M. Morgenthau investigated the case for possible abuses in justice. DNA evidence showed that Trisha was raped and beaten by Matias Reyes, a convicted rapist serving thirty-three years to life, who confessed to acting alone. He could not be charged for the crime because the statute of limitations had expired. The five teenagers had been in the park, coincidentally near the rape scene, not knowing that a rape was happening. Years later, after their exoneration, the men acknowledged being in the park, and that they had committed some unrelated assaults, robbings, and beatings. There were several gangs roaming that night, sometimes joining up and sometimes splitting. They admitted to knocking down a man and dragging him into bushes where they poured beer over him. They admitted to eight attacks on people in the park.

For Trisha, life was interrupted by that coincidental night. It was a life that had another coincidental turn. Salomon Brothers is no more, and Trisha is a different person. "I went on a run," she wrote in her memoir, "and had my life interrupted. No one comes that close to death without being transformed in some way, and I've learned to accept the changes, both positive and negative." In 2004 she wrote:

> I'm not sure why this is so. In the intervening years, there have unfortunately been innumerable beatings

> and countless rapes (during the week I was attacked, twenty-eight other rapes were reported across the city), yet my case is remembered while the others are forgotten by all but the victims, the victims' immediate families and friends. Perhaps it is because this assault revealed the basest depravity human beings are capable of—the attack was believed to have been committed by a group of teenagers between the ages of fourteen and sixteen, out only to have some "fun"—and people shuddered to realize such cruelty exists in our exalted species.[18]

There is a serious need for the general public, from which we get our jury pools, to be informed about how DNA works and how flukes happen, even in the most carefully conducted police investigations. A sneeze can bring an innocent person's DNA miles away by train or plane, or by just a leaf blowing in the wind. Even fish can come to a newly created pond by eggs sticking to a bird's webfeet. The public needs to understand close matches and methodology, how short pieces of DNA strings can have coincidental repeats with no apparent physiological function, and how conclusions are drawn from random coincidental possibilities of hair matching, shoeprints, fingerprints, voice, and, yes, eyewitness misidentification.

Complete understandings of the sequencings of the four nucleotides that make up the DNA are not so important, but knowledge of the ease of contamination, and that copies of nucleotide pairs are rare in some populations and more common in others, can mean a great deal to the judicial fate of a suspect.

The truth of evidence (guilt or innocence) can be affected by hidden coincidences, and that the public should never make any judgments of guilt or innocence simply based on DNA profiling or eyewitness identification alone. The hope here is to create a public understanding of the complexity, so media and jurors understand that the criminal evidence, no matter how scientifically it is explained, is not always as true as it is portrayed in the courtroom.

The five accused teenagers confessed to the crime after their arrest.

Why—you are asking—would an innocent person confess to a crime he or she did not commit? There is a serious misconception of prosecution accuracy boosted by television and movie renderings of American criminal justice. First, we must understand that there are 2.2 million people in American prisons and well over 2 million of them are there because they have accepted a plea bargain to avoid the gamble of a jury trial that might dictate a maximum sentence. For the more heinous crimes, such as rape and murder, the gamble is between life in prison or death. So, the accused makes a risk-management, cost-benefit judgment when confessing to a crime that he or she did not commit. It is a natural self-defensive option, a rational decision brought on by the pressures of an imperfect criminal justice system. Imperfect, because a plea bargain almost always vindicates guilt and the gamble is always biased on the side of the prosecution. We might think that few of the accused and innocent would confess, but the Innocence Project reports that 10 percent of the accused pleaded guilty to crimes they had not committed, and that in about 30 percent of DNA exoneration cases the accused signed confessions. Many of the accused are under duress and coercion, ignorant of the law, misunderstanding what they are signing and, most often, feel that they are avoiding a harsher sentence. The Central Park five were children, manipulated, pressured by false advice that they could "go home" as soon as they admitted guilt.

A confession through plea-bargaining gives a poor person with limited resources and other troubles a way of getting a smaller sentence. In the words of Jed S. Rakoff, the United States district judge for the Southern District of New York, "Every criminal defense lawyer . . . has had the experience of a client who first tells his lawyer he is innocent and then, when confronted with a preview of the government's proof, says he is guilty. . . . But sometimes the situation is reversed, and the client now lies to his lawyer by saying he is guilty when in fact he is not, because he has decided to 'take the fall.' . . . Rarely, however, do [Americans] contemplate the possibility that the defendant may be totally innocent of any charge but is being coerced into pleading to a lesser offense because the consequences of going to trial and losing are too severe to take the risk."[19]

Exonerations for the Innocent

The United States has the largest prison population in the world, weighing in with just under one quarter of the whole world's prison population.[20] Most incarcerations are for nonviolent offences. At the time of this writing approximately 2.3 million persons are being held in federal and state prisons in the United States, more than 840,000 of whom (nearly 37%) are African American. That's a 546 percent increase since 1970 and an unsustainable increase of more than 50 percent in just the last 6 years![21] It means that 1 in 100 American adults are behind bars, leaving 1 child in 28 with a parent behind bars at a staggering cost of $260 billion dollars a year.[22] An inhumane madness that wastes human potential! Some people believe that mass incarceration is the cause of the dramatic decrease in the crime rate. (Since its peak in 1991, the rate of violent crime has decreased by 51%, and property crime by 57%.) What sounds logical is not always valid. The causes are not so apparent. Coincidence or fluke, we know that there are hidden variables in the hundreds that might account for the dramatic decrease in crime. A recent extensive, rigorous, and sophisticated empirical analysis research report from the Brennan Center for Justice, using the most recent comprehensive dataset to date, concludes that "At today's high incarceration rates, continuing to incarcerate more people has almost no effect on reducing crime."[23] This 140-page study is impressive in its scope that uses a mathematical method that distinguishes the effects of each variable in comparison with others. That's fine for establishing a correlation, but just a mere wink at causality.

Surely, we know that there are causes, but we cannot confidently know them. So, surely, we cannot say for sure that increased incarceration leads to a decrease in crime. Incarceration contributes greatly to family breakups; that innocent children are psychologically harmed; and that without intensive rehabilitation, the ex-con will find it hard to learn how to become an employable contributor to society. What we can say for sure is that the United States leads the world in per capita documented incarceration rates, behind Russia and Rwanda. It has a higher percentage of incarcerated people than has any other democracy in the world, with one-quarter of the world's total prison population. In 2014, 515 of the 1,409 exonerations in the United States were of prisoners on death row. That's a staggering rate of 16.8 percent![24] Since 1976 there have been 1,386 executions in the United States and just 144 exonerations of death row verdicts.[25] That means that since 1976 almost 1 in 10 people should not have been sent to death row.

The US Supreme Court expressed its moral justification of capital punishment with a claim that capital punishment is permissible in an advanced society as long as there are procedural safeguards in place that reduce the risk of execution of innocent people.[26] The key word in that last sentence is: *reduce*. But the risk of executing an innocent individual cannot be eliminated entirely. So, if we accept Maimonides' maxim as expressed in the epigraph of this chapter, it seems clear that capital punishment should be abolished. Justice John Paul Stevens came to this conclusion in 2008 when he said that the court's justification of the death penalty is not "tolerable in a civilized society."[27] No matter how the argument is framed, the issue is not a logic-tight chain of inferential scientific arguments. There will always be false positives and false negatives; there will always be innocent people put to death and guilty people set free. The variables of behavior and nature are too many and too complex to tie down to human decisions that might or might not be guided by fact. No legal system is likely to be able to eliminate the risk of executing innocent people. In August 2014 there were 3,070 inmates on death row in the United States.[28] One recent study estimates that close to 123 of them may have been wrongfully convicted.[29]

I accept Maimonides' maxim. And I agree with John Paul Stevens' belief that it is not likely that we will ever eliminate the risk of executing innocent people. Yet I would go further to say comfortably that for the foreseeable future it will be impossible to eliminate the risk. Why? Because we are dealing with billions of variables that depend on surrounding circumstances mixed with human nature mixed with the phenomenally complex electro-chemical performance of a thick soup of neurons acting in a billion variable environment.

Research from the Innocence Project in 2009 found that in 239 convictions that ended in DNA exoneration, 179 were first convicted on eyewitness misidentification.[30] By 2013 the number of DNA exonerations had risen to 250.[31] In 114 cases, the true guilty perpetrator (by alleged DNA evidence) committed violent crimes while the innocently convicted person was serving time in prison.[32] As of this writing, in the United States there have been 1,587 exonerations in the last fifty years.[33] Almost every day we read about another case. We learn about people being accused by detained reluctant witnesses, sometimes in police inquiry rooms and sometimes in hotel rooms. We learn that they are held until they agree to testify. We learn that prosecution lawyers are advised to not take notes when their witnesses give inconsistent statements so as to avoid potentially exculpatory evidence.[34] We learn of police errors and prosecutors' misconduct. We learn of evidences proving definite innocence that are never turned over to defense lawyers. We learn of confessions handwritten by police after interrogations of suspects without lawyers present. We learn of convictions that have no physical evidence linked to crimes. And we wonder whether the Constitution has the moral right to permit capital punishment. Maimonides saw the problem back in the Middle Ages. His moral dictum, "It is better and more satisfactory to acquit a thousand guilty persons than to put a single innocent one to death," is as wise today as it was back then.[35]

Chapter 12

Discovery

Dans les champs de l'observation le hasard ne favorise que les esprits préparés. (In the fields of observation, chance favors only the prepared mind.)

—LOUIS PASTEUR[1]

GREAT INVENTIONS AND discoveries may be boosted by the proverbial *aha*. But sometimes that aha is boosted by something that either goes wrong or something that just happens through no apparent cause—some kind of interference from an ingredient in the lab that was part of a different experiment, or a tool that had a timely debut on the market, or something that simply goes wrong in the experiment.

Chemists had been working with molecular bonds for centuries before knowing anything about why or how those bonds worked. Before the twentieth century they knew nothing about shared electrons, because they didn't know about electrons. And yet they were able to do fabulous chemistry by knowing how atoms and molecules interact and transform to create new compounds. They were able to analyze the reactions of molecules and transformations under heat and light, and even to prepare complex compounds that included polymers and metal alloys, without ever understanding the critical role that electrons played in creating the necessary bonds. They understood that gases always react in a balanced relationship of proportions with one another. All without knowing that electrons had something critical to do with the reactions and bonds.

These were the scientific discoveries of unusual people, who by some unaccountable luck met timely flukes and coincidences and wisely recognized them as clues to the answers of big questions. They show us that unplanned happenings can be as useful to discovery as purposeful hypotheses. They show us that accidents in scientific observations can shape the way we think about what we see and change the world for the better. There are many such stories, including how some accidental dyes of William Perkins contributed to understanding immunology and chemotherapy; the discovery of penicillin through the work of Alexander Fleming, Howard Florey, and Ernst Chain, in whose untidy lab a staphylococci culture was contaminated by a fungus that happened to surround and destroy the staphylococci. Consider, too, the story of Alan Turing, Ralph Tester, and other WWII code breakers of Bletchley Park whose breaking of the "unbreakable" Enigma code played a significant role in which side would win the war. These were very clever gifted people, but thanks to a bit of good luck from a few German coding errors, the English cryptographers were able to work out the logic of Germany's coding machines. The knowledge gained not only helped the Allies win the war, but also helped in the invention of the world's first partially programmable computers.

In 1869 Dmitri Mendeleev had a dream in which he claimed to have arranged the elements in a table according to their atomic weights.[2] He awoke the next morning to put together the periodic table. It was a time when national weather agencies were beginning to collect data on temperature, precipitation, and any other climate data that could be trusted. In those years, chemistry was not about the atom. Chemistry had already been given its scientific roots almost a hundred years before, when Antoine Lavoisier discovered the role of oxygen in combustion and established that mass is always conserved. Still, in 1869, when Mendeleev first published his periodic table, chemistry was flying blind in its experiments, knowing nothing about the internal mechanisms of the atom. It was a simple time when railways connected cities all over Europe and Russia, though it was still not so easy to travel between countries. And Saint Petersburg, a city of white nights, where Mendeleev lived and taught, a city of high fashion, wealthy aristocrats, and exciting entertainment, was also a severely overcrowded unhealthy place with bad water, malnutrition, poor sanitation, and dispersing, festering diseases.[3] That same year Swiss physician Friedrich Miescher isolated DNA from the pus of used surgical bandages. Miescher, also flying blind, did not know at that time that it was a hereditary molecule encoding genetic instructions, but it did pave the way for understanding that DNA is the carrier of inheritance.

It was right about that time when many physicists were experimenting with Crookes tubes, blown glass tubes under partial vacuum with electrode connections inside at each end. The experiments were trying to understand the glows inside the tubes. We now know what happens when a high voltage is applied to a Crookes tube containing rarefied gases: A small number of charged gas molecules (positive ions) in search of electrons are excited and collide with other gas molecules, knocking off some electrons to create more positive ions. The positive ions are then attracted to the negative electrical terminal. As they hit the surface of the metal terminal, they knock off a large number of electrons. Attracted to the positive terminal, they move in and across the tube, forming a glowing ray of electrons, a *cathode ray*. During more than thirty years of experiments, scientists were playing with different gases, without any deep understanding of what was really going on. They knew nothing about negatively charged particles, those electrons within the gas atoms. And they knew nothing about what was causing the light itself. New information about what was going on came from flukes or coincidences that they did not understand. One glass glowed red, another glowed green. There was little fundamental understanding of why. For instance, they didn't know that in the partial vacuum many electrons of very low mass and direct paths to the positive terminal were being attracted with a built-up momentum and speed. The closer those electrons got to the positive terminal, the greater the attraction. We now know that those electrons, heading towards the positive terminal, would gain speeds relatively close to the speed of light. Some would fly right past the positive terminal and hit the glass atoms of the tube to knock their orbiting electrons into higher energy levels for an instant before falling back to their original energy levels. In falling back, elementary particles of light (photons) would be emitted, and so the glass would glow a kind of greenish-yellow luminescence.

X-ray fluorescence, the emission of *light* by *electromagnetic radiation*, is a bit more complicated. Wilhelm Conrad Röntgen discovered X-rays by accident when experimenting with electric currents in a glass container under partial vacuum. A screen coated with barium platino-cyanide (a fluorescent material) just happened to be set up in his laboratory for a different experiment. Had that screen not been there, who knows how many people would have shorter lives from the delayed discovery of X-rays and their uses. Röntgen was not looking at the screen that was some distance away. Not expecting it to have anything to do with his experiment, he caught a glimpse of the screen in the corner of an eye. Something happened that seemed to be independent of his experiment. It was a fluke, but a fluke of many consequences.

Let's look around Röntgen's lab at the University of Würzburg as it was on November 8, 1895.[4] A large window looks over a narrow avenue of Norway maple trees that have lost most of their leaves. Spindly mahogany tables of different heights line a wall in the light of the window. A clutter of instruments, metals, motors, flasks of all shapes, and coils are on the tables. There is a pendulum clock on the wall alongside a shelf of hanging wires of different lengths. Glass tubes precariously lean against each other on one table. An electric light fixture with a clear incandescent bulb hangs from the ceiling connected by a low-hanging wire to an outlet near the clock on the wall. The rest of the room is almost empty. Aside from its brightness from the outside light, it looks no different than almost any other nineteenth-century chemistry lab. There is no curtain on the window.

The man in the lab is Röntgen. He is fifty. His hair is thick and black. His beard is long and black, starting to gray. Since the beginning of 1895 he has been experimenting with electricity by firing electrostatic charges though partially vacuumed glass tubes. On November 8 he is experimenting with cathode rays that created a visible glow in glass containers. The rays are not visible outside of a partial vacuum, so the natural question is on his mind: could some of those invisible rays escape from the glass container?[5] In an attempt to block transference of rays or to spot escaping rays across the room, he covers the container with a cardboard shield and darkens the room. Across the room the screen glows, and by controlling the vacuum and current in the glass container he can control the screen's glow. The glow is faint. In experiment after experiment, the result is the same. Even moving the screen farther away, the result is still the same. When the laboratory is totally darkened, the result is the same. The glass container is shielded further, and the result is exactly the same. The wavering light on the screen cannot have been a result of any light other than the cathode rays coming from the electric current in the glass container. It means that the rays were passing through the shield and zooming through the air to hit and illuminate the screen. It is a new kind of ray, never before discovered, an unknown ray.

Since x had been used to designate the unknown in mathematics ever since Descartes introduced it, Röntgen decided to call these new rays "X-rays." James Clerk Maxwell and Michael Faraday had already predicted the existence of invisible electromagnetic waves that could travel through free space over some distance. Three years before Röntgen discovered X-rays, Heinrich Hertz was experimenting and demonstrating that cathode rays could penetrate thin metal foils. Meanwhile, Hermann von Helmholz was developing mathematical equations for theoretical X-rays, hypothesizing that real X-rays did exist and could travel at the speed of light.

Imagine Röntgen's surprise when he tried to block the rays by placing his hand between the container and the screen, and saw the bones of his hand on the screen, a skeletal picture! He was peeking into his own body. From the biographies that were written long after his death, we learn that he had no intention of placing a body part between the container and the screen.[6] It just happened. Very likely he was the first person ever to do so. He tried to block the rays with other material objects—wood, metal, paper, rubber, books, cloth, platinum, and all sorts of household objects. Some objects allowed rays to freely pass through them; others blocked rays. A photograph of a wooden spool of wire showed only the wire, with only a faint shadow of the spool. In a follow-up experiment, Röntgen tested the X-ray transparency of aluminum plates of thickness 0.0299 mm, by piling on sheet after sheet. He could not discern much difference in transparency between one and thirty-one sheets, and small distances from the barium platino-cyanide screen also did not make much of a difference. X-rays could pass through living tissue unobstructed, but not through bones or some metals, such as lead. They could pass through wood, but not through coins. Röntgen soon had the bright idea of replacing the screen with a photographic plate. He beamed X-rays through a closed wooden box with a coin inside to capture a clear photograph of just the coin, as if the box were not there at all. He photographed the hand of his wife, Bertha. She could see her finger bones and the ring she was wearing. The photo became quite famous after a Viennese newspaper printed it.[7] It was probably the first photo ever taken of the inside of a living hand. For some, it was a curious phenomenon, and for others it was a joke. Daily, weekly, and monthly presses were active with stories of the new photography. *Life* magazine published a cartoon lampooning the new kind of photography that led the imagination to extremes.

A satirical poem was in the next issue of *Life*.[8]

She is so tall, so slender, and her bones—
Those frail phosphates, those carbonates of lime—
Are well produced by cathode rays sublime,
By oscillations, amperes and by ohms.
Her dorsal vertebrae are not concealed
By epidermis, but are well revealed.

Barbara Goldsmith, in her book *Obsessive Genius*, writes, "As X-rays swept the world they soon became the subject of cartoons—husbands spying on wives by X-raying through locked doors, X-ray opera glasses that revealed naked bodies under the costumes. . . . A London firm sold X-ray proof suits."[9]

Great scientific discoveries have scientific forefathers, some more than others. They hardly ever happen by direct shots. Most take repeated attempts, and some come to success because of a simple fluke that happens along the way. They may come accidentally by starting out through accident, but almost always—or maybe always—they have followed clear clues aimed by some conjectured or known theory. That is why there is no reason to suspect that Röntgen's discovery would not have happened had the barium platino-cyanide screen had not been there. Other physicists were studying the effects of cathode rays, and it is safe to say that research in that area at the end of the nineteenth century was very exciting. The English physicist William Crookes (for whom the blown glass partial vacuum tubes were named) was able to produce a beam of radiation coming from the cathode to thereby discover cathode rays, and launch a frenzy of cathode-ray research. By using concave cathodes to focus cathode rays, Crookes was able to produce enough energy to produce some X-rays, though losing much of the energy as heat. He thought it strange that some unexposed photographic plates stored nearby were fogging. Without thinking further, he returned the plates to their manufacturer, claiming them to be defective.[10] And in 1888 Philipp Lenard used cathode-ray tubes in experiments with high-frequency ultraviolet radiation. If only he had a low enough vacuum within the tube, and generated a higher voltage, he would have generated enough X-rays to detect fluorescence outside and beyond the quartz end of his tube. But the vacuum pressure was not low enough and the voltage was not high enough. So, he never detected any of the X-rays he generated.

Michael Faraday considered fluorescence when, in 1838, he began working with electrical potentials through partially evacuated glass tubes. Afterward, young German physicists experimented with partially evacuated glass tubes of all sorts and shapes. They tried neon, argon, and even mercury vapor under high voltages. German physicist Heinrich Geissler started putting metal electrodes into partially evacuated glass-blown cylinders in 1857 to show glowing lights. Yet, in all those years, with all those astute scientists working in relatively well equipped university labs virtually similar to Röntgen's, the accidental notice of an action at a distance, a shimmering faint light a short distance from the tube, an X-ray, was not spotted. They did not detect electromagnetic radiation of such short wavelength that could produce any faint shimmering of light outside their glass tubes.

We will never know how close we came to a delay in the discovery of X-rays, and we can only guess (because the data is too skewed to give evidence) that in the last twelve decades since Röntgen's discovery, "X-rays have saved more lives than bullets have destroyed."[11] Had the discovery not happened when it did, it is entirely possible that the internal nature of the atom would not have been discovered for at least another decade, and that lack of knowledge would have delayed the great discoveries that were to come in a chain leading up to all the vast changes in the world map as we know them today. The actual Röntgen discovery has been told and retold. Röntgen gave few interviews. One of the most respected accounts comes from H. J. W. Dam, a *McClure's Magazine* science reporter.[12] It is a beautiful piece to read, full of detail and descriptions of Röntgen, his laboratory, and his experiment:

> "Now, Professor," said I, "will you tell me the history of the discovery?"
>
> "There is no history," he said. "I have been for a long time interested in the problem of the cathode rays from a vacuum tube as studied by Hertz and Lenard. I had followed theirs and other researches with great interest, and determined, as soon as I had the time, to make some researches of my own. This time I found at the close of last October. I had been at work for some days when I discovered something new."
>
> "What was the date?"
>
> "The eighth of November."
>
> "And what was the discovery?"
>
> "I was working with a Crookes tube covered by a shield of black cardboard. A piece of barium platino-cyanide paper lay on the bench there. I had been passing a current through the tube, and I noticed a peculiar black line across the paper."
>
> "What of that?"
>
> "The effect was one which could only be produced, in ordinary parlance, by the passage of light. No light could come from the tube, because the shield

which covered it was impervious to any light known, even that of the electric arc."

"And what did you think?"

"I did not think; I investigated. I assumed that the effect must have come from the tube, since its character indicated that it could come from nowhere else. I tested it. In a few minutes there was no doubt about it. Rays were coming from the tube which had a luminescent effect upon the paper. I tried it successfully at greater and greater distances, even at two metres. It seemed at first a new kind of invisible light. It was clearly something new, something unrecorded."

"Is it light?"

"No."

"Is it electricity?"

"Not in any known form."

"What is it?"

"I don't know."

And the discoverer of the X rays thus stated as calmly his ignorance of their essence as has everybody else who has written on the phenomena thus far.

Other accounts bring in clear references to the barium platino-cyanide–coated paper that just happened to be on a table at some distance in the room and to the accident of the discovery. In other secondhand reports, the barium platino-cyanide screen was on the table because Röntgen was thinking it was more efficient than other fluorescent coatings.[13] At his 1896 Wurzbürg Physical Medical Society lecture he talked about how he first observed the fluorescence of the barium platino-cyanide paper, how he had discovered that the fluorescence appeared only when a charge went through the covered Crookes tube, and how that same phenomenon happened even when the fluorescent coated paper was placed at a farther distance.[14] And then he said, "I

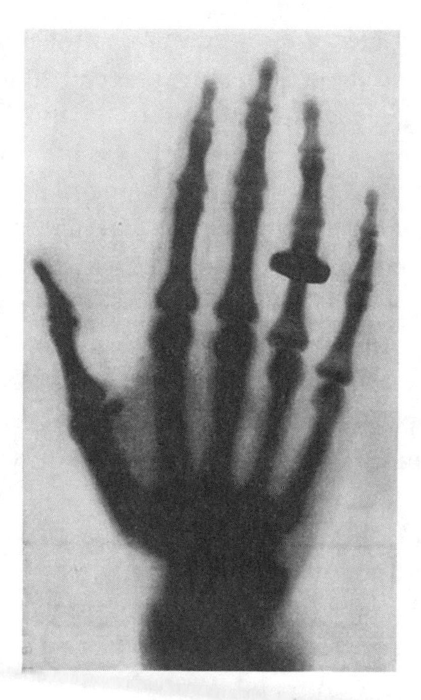

Figure 12.1. Röntgen's X-ray shadowgraph of a lady's hand showing the bones, fingernails, and ring.

found by accident that the rays penetrated the black paper. I then used wood, paper, books, but still I believed I was the victim of deception. Finally, I used photography and the experiment was successfully culminated."[15] Photos, such as the one appearing in Figure 12.1, circulated in newspapers around the world on December 22, 1895.

Soon after, the idea was applied to medical science, enabling physicians to peek inside the human body for tumors, abscesses, cavities, bone structures, and so on, which could not be discovered by conventional means. It is not clear that Röntgen had any proficient sense of the extent of the value his technique would have on medical diagnoses of internal maladies.

He had every intention of returning to the intended experiments that involved the screen, but got so caught up in his X-ray experiments that he never did return to the initial experiments involving the screen.

As the nineteenth century was coming to a close, scientists still knew almost nothing about the internal structure of the atom. Scientists had known about electricity for centuries. They knew how to generate it. By 1880 incandescent bulbs of one kind or another had been lighting streets in London, Paris, Moscow, and the United States. Scientists even knew that forces and energy pervaded all of space. And from Faraday and Maxwell they knew about electromagnetic wave theory. But electrons had only been discovered in 1897, breaking the ancient notion that the atom was the smallest part of anything. How electric currents carried in wires from one point to another was still a mystery. The success of chemistry in the face of such a mystery is quite amazing, considering chemistry had been well established a century before. And though cathode rays and X-rays were well established in theory, nobody at that time had ever actually "shown" their existence. The verb "to show" as used in that last sentence would not have necessarily meant visible using some instrument such as a microscope. Science had plenty of examples of scientific phenomena that could not be seen by instruments. And nobody at that time knew how fluorescent streams of electricity got from one terminal of a Crookes tube to the other.

J. J. Thomson's 1897 experiments with cathode rays showed that the rays themselves were not atoms flowing from one terminal to the other; rather, they were material components of atoms. Atoms were no longer just solid balls that had no parts. Protons and electrons were predicted to exist, because, though they could never be seen, they could be measured by their effect on instruments. In a 1934 interview, Thomson rhetorically asked, "Could anything at first sight seem more impractical than a body, which is so small that its mass is an insignificant fraction of the mass of an atom of hydrogen, which itself is so small that a crowd of these atoms equal in number to the population of the whole world would be too small to have been detected by any means then known to science?"[16] In the next few decades, science would come from knowing next to nothing about the atom and absolutely nothing about electrons and protons, to an understanding of some of the deepest secretes of the material universe and of the internal mechanisms of the atom. By 1939 it would discover nuclear fission, yet even today the basic building blocks of the atomic nucleus remain enigmatic, consisting of particles bewilderingly called "up quarks" and "down quarks," each a pulsating mass of even smaller parts, all held together by a strong force.

There are many classic accidents of scientific discovery in the popular history of science: the discovery of the antimalarial drug quinine by a South American Indian suffering from malaria who drank water near a cinchona tree; how insulin was discovered from noticing flies on the removed pancreas of a dog; and stories of Descartes inventing his coordinate geometry by lying in bed, watching a fly. There are many other stories involving chemical inventions that are more technological inventions than basic science discoveries. They deserve honorable mention, but are not included here for the simple reason compactly expressed by Louis Pasteur as "Le hazard ne favorise que les esprits préparés."[17] Moreover, many of those stories are told out of the context of a scientist's original notes. Exaggeration easily finds its way into a story by wishful embellishment. It is the natural backdrop of storytelling. Before anything is accomplished, there is always an accumulation of primary essential work. Dig into the actual history of a discovery and you will almost always find the discoverer seeing from the collective shoulders of giants. Even that famous line of Isaac Newton's, "If I have seen further it is by standing on ye shoulders of Giants," was not original. Newton did indeed write this in a letter to Robert Hooke in 1676.[18] Its originator was twelfth-century French Neoplatonist philosopher

Bernard of Chartres, who compared his generation "to [puny] dwarfs perched on the shoulders of giants." Bernard pointed out that we see more and farther than our predecessors, not because we have keener vision or greater height, but because "we are lifted up and borne aloft on their gigantic stature."[19] Surely there are some who might stand on the shoulders of giants and not see far, and there are some who might not need giants because they stand on the collective shoulders of lots of ordinary people with a dedicated purpose. I prefer Steven Weinberg's recognition of giants. In his excellent book of essays on modern physics and science policy, *Lake Views*, he writes, "We recognize that our most important scientific forerunners were not prophets whose writings must be studied as infallible guides—they were simply great men and women who prepared the ground for the better understandings we have now achieved."[20]

Mold may well have been on a petri dish in Alexander Fleming's lab, but the fact that it was there in the first place makes me suspect that there was some connecting purpose. It did not grow there on a piece of humid bread as some folklore accounts have it. It was in a petri dish! Connecting purposes guide scientific discovery. As with monkeys trying to write a line of Shakespeare, haphazard aims almost always miss their targets.

Chapter 13

Risk

LUCK RARELY COMES without risking the possibility of loss in a universe of opposing chances. Playing the equities markets is a game, just like poker, where you compute the probabilities of being dealt a promising hand, weigh the risk of not getting such a hand with what could happen if you lost the pot, and weigh the odds of your hand overriding the one you are trying to beat. That is how it is in the financial markets. You compare the amount of risk you are willing to take with the return you might clear. You buy and sell a stock according to appraisal and judgment, looking at its past and current earnings, its growth potential, and its competition. You look at the balance sheet. In the end, no matter what, your investment, as shrewd as it is, is still a risk. More than anything, it is wishful thinking.

You may be thinking that those financial engineering sharps, those hedge fund quants who use quantitative analysis to steer their way through bull and bear markets, would know how to profit in the chase. They play the game of finance very astutely, but it still comes down to wishful thinking. They make money by chasing the volatility of stocks driven by small-time investors who buy and lose. Maybe that is all okay. But when financial institutions buy and sell in huge volumes, their transactions can drive strong, resonant waves that could bring down the whole world's economy.

The market place is now almost entirely global: weather changes in the Pacific can affect grain markets in Chicago; droughts in the US Midwest can affect sales of farm equipment in Canada; floods in Mississippi can deplete timber forests in Brazil. Weather conflagrations are at the center of risk odds. It does not take more than one person, prone to risky behavior and oblivious to looming consequences, to fiscally rock the world.

Take the story of the French multinational banking and financial company Société Générale, now 150 years old. Had the US government not bailed out AIG, the insurance giant that insured Société Générale, the company would probably not have made it through its 144th year.

Between January 2005 and July 2008 a thirty-one-year-old French trader perpetrated the largest trading fraud in history. Jérôme Kerviel cost Société Générale a staggering net loss of 4.9 billion euros by selling short 10 million euros of a European insurance company, hoping that its stock price would fall. It was an astonishingly lucky gamble. There were no hints that the price would fall, but by Kerviel's fluky luck the entire London FTSE fell. Kerviel could not have known in advance that Islamist suicide bombers would explode themselves during rush hour aboard three London Underground trains and on a bus, killing fifty-two people and injuring seven hundred. He made a profit of half a million euros. His win contributed to a "favorable history of reinforcement."[1] Kerviel told the police, "It makes you want to continue; there's a snowball effect."[2] So, his risky behavior intensified with covert purchases of hundreds of millions of euros. Amazingly, those, too, turned into substantial profits.

Kerviel had one problem. To not draw attention, he had to conceal his purchases by trading off the books to offset his gains. Smartly thinking that the global markets would suffer severely from the subprime real estate mess, he began selling short millions. It wasn't long before he found himself selling short billions. It was a risky gamble in hopes that the subprime mess would bring the markets further downward. That is exactly what did happen. By the end of 2007 Kerviel's activities gained a colossal 1.5 billion euros.

Then came his big, big, big mistake. By the beginning of 2008 he started betting on futures, extending his exposure to nearly 50 billion euros. He was thinking that the market had hit bottom and, like all market cycles of his lifetime, recovery was inevitable.

That's when things began to go very badly. The equity markets were falling, leaving Kerviel's futures hugely exposed with no hedge to cover it. An exposure the size of 50 billion euros could have bankrupted Société Générale.

Blindsided by the loss, the bank was forced to sell off futures. How do you sell off 50 billion euros without anyone noticing? A sell-off of that size could cause a panic. You cannot. (In England, in these days after 9/11, an ordinary bank customer cannot transfer more than £5,000 at any one time to a different account outside the UK.) Though the bank had to experience a huge loss that was far less than 50 billion euros, it hadn't exposed itself to the true amount of exposure. It sheepishly sold off 6.4 billion euros in "the largest single day trading loss for a single company in banking history."[3]

Clearly, the London Underground bombings played a major role in the chain of events that led to Société Générale's loss. But Kerviel could not have known in advance that he would make such a huge killing by selling short 10 million euros of a European insurance company. The bombing was a coincidence that had no a priori connection to Kerviel's plans. It made him wealthy. The continuing equities tumble brought him down. Had the market truly hit bottom when he began betting on futures, things might have been different. He and the bank might have gotten away with the deception of trading on the bank's account without authorization, and nobody would have known about his enormous risk. Did the risk managers ignore Kerviel's suspicious transactions, or was it just a colossal fluke that they missed the passing of a few billion euros? "I find it hard to believe," Hélyette Geman, a professor of mathematical finance at the University of London told the *New York Times*, "that the risk management systems and all the auditors did not indicate anything at any level."[4] In the end it is all about greed. Where there is money, there is greed.

But what's a billion euros? As Joseph Mirachi captions in his famous 1975 *New Yorker* cartoon illustrating two generals presumably discussing a military budget, "You blow a billion here, you blow a billion there. It adds up." Talk to Nick Leeson, the rogue derivatives trader who, back in 1995, brought down Barings Bank—the oldest investment bank in England—by gambling in the futures market and losing £850 million ($1.3 billion). His unsupervised and unauthorized speculations might have gone very well had it not been for the Kobe earthquake. It was Leeson's high-stakes poker game, lost by a coincidence of monumental proportions. Leeson was playing poker with short-term futures on the Singapore and Tokyo stock exchanges, betting that the Japanese stock market would be secure. However, early the next morning (January 17) the Kobe earthquake hit, sending Asian markets into a tailspin. Chasing his losses, he made a series of increasingly risky investments, betting that the Nikkei Stock Average would recover. It didn't. Like many gamblers who chase their losses, he continued to sink into deeper trouble.[5]

In the twentieth century, cases of risky Wall Street leverages were isolated and insulated from global effects. In this century, economic globalization has changed all that; almost all banks are entwined in a strong web of dealings that make almost all banks vulnerable to the behavior of one. In the three days while Société Générale was frantically liquidating Kerviel's futures, other traders were making money selling short, and betting on a falling market. When world markets fall, some people still make money. The money doesn't just disappear. Bank collateral can even increase through government handouts.[6]

Flukes of Market Disruption

Market responses to earthly disasters, such as tsunamis and earthquakes, as well as terrorist attacks, wars, and Ebola epidemics are not coincidences. They have very apparent causes associated with diminished market circumstances—parts and materials supply disruption, weakened buying power, and market jitters, just to name a few. But most natural megadisasters are not scientifically predicted, and those that are, have come as fast as lightning strikes to catch the market unaware. Earthquakes are not coincidences. They have a definite cause. But their time of happening is almost always coincidental. As a leading current textbook on seismology put it,[7]

> [We have] no ability to predict earthquakes on time scales shorter than a hundred years, and only rudimentary methods to estimate earthquake hazards. . . . Our best response seems to be to show humility in face of the complexity of nature, recognize what we presently know and what we do not, use statistical techniques to assess what we can say with differing degrees of confidence from the data, and develop new data and techniques to do better.

Similarly, mathematician Florin Diacu tells us this in his excellent book *Megadisasters*:[8]

> Like many other sciences, seismology uses mathematical models to examine how earthquakes occur and develop. The rupture triggered during an earthquake involves several physical actions, which lead to the

> propagation of various waves through the Earth's crust. Since most of these processes can be only guessed at, the models are simpler than the physical reality.

The eminent disasters of tsunamis are somewhat predictable on time scales of a few hours, but only after they actually occurred far out at sea. Intelligence agencies sometimes have advance information of imminent terrorist attacks, but not always. The attackers and their commanders know the places and times, but successful attacks happen at times and places that surprise us.

We have drawn our attention to just a few of the unforeseeable, yet possible, megadisasters. There are others, and there will be others, that we cannot yet imagine. Like all gambling, they keep us as vigilant as we were a hundred thousand years ago when we were still in caves, alive, waiting for the chance to bravely come out for the hunt, not knowing what would befall us from the earth and sky. That, too, was a market time, the ultimate market time, the essential gambling time, filled with unknown circumstances that had no apparent causes and with the will and excitement of just being and staying alive.

Coincidence events are generally unexpected because they seem to be extremely rare, but they *should be* accounted for in risk assessment *because* they have not happened in a long time. Any such consequence is predictable because of two competing mathematical models. One tells us that there is a tendency for outcomes to cluster close to the mathematically predicted mean, and the other is a principle of probability that tells us surprising things actually are likely to happen with large enough sample sizes. On the surface, we view the outcomes of most events by centering our sights and calculations on a small number of possibilities. Such centering ignores unexpected catastrophic events because they seem to have remarkably low probabilities of occurring. In reality, those probabilities are far higher than we think.

This accounts for why observed success ratios are more likely to approach mathematically computed probabilities over the long run. However, along the way, unforeseeable coincidences of natural phenomena can create short-lived volatile success ratios. Surprisingly, short-lived vulnerabilities can tilt the risk of long-term successes enough to disturb the mathematical predictions of success.

Most games of chance have reasonably accurate computable odds of winning. Their probability models are based on game structures, not on external connections with unaccountable natural phenomena. The best gambling strategies discount quantifiable risk of unforeseeable coincidences. Financial markets, on the other hand, are not exactly structured games of chance.

Traders often willingly ignore the possibility that some small rare event may cause a global catastrophe. They gamble on the market, believing that it runs by some kind of perfectly efficient rule, when, in effect, it is no more predictable than the law of large numbers' ultimate prediction of the outcome of a flipped coin. The trader is supposed to study the news, investigate IGPs, analyze futures, assess liabilities and faults, assess leaders and connections with other companies, and evaluate histories. Few traders study any global consequences of catastrophic possibilities.

Today's commercial markets are so tightly connected that the failure of one risky business often leads to the failure of others in the linkage. We can no longer count samples as if they are independent, like what we do when we are looking at coin flips, die tosses, or roulette spins.

It doesn't take much stock market volatility to give consumers shivers. When the market takes one of its rare sharp turns, perhaps spooked by a discouraging event, such as a near collapse of one of the most respected banks in the world, it's possible to skid off track. The daily fluctuations in value of any single company affect that of a pack of others; how could one handicapper know what will happen in a world of daily political, social, or economic events? Hurricanes pass by offshore oilrigs, autoworkers strike to hold on to their benefits, juries grant huge class action damage awards against pharmaceutical firms, orange groves freeze, CEOs are indicted for fraud (or should be), Ebola virus panics airline passengers, and so forth. Who can say whether those events are flukes of timing? A rare sharp turn in the Dow 30 spooked by any discouraging shake-up event, such as a near collapse of a giant bank, could perturb the market enough to flick it out of its moderate equilibrium. When the value of any single large company volatilely fluctuates there is a ripple effect. Any one of a large number of unforeseen events with unexpected outcomes seemingly caused by unanticipated coincidences could throw the marketplace one way or another.

How should we factor in some unexpected outcome caused by some unanticipated coincidence? Sometimes there are warnings we are able to recognize, as happened in Haicheng in 1975, when the Chinese experts saw the precursors, identified foreshocks, and understood the animal behavior in the surrounding countryside, and correctly predicted the time of the next quake. It was a fluke. The Haicheng warning seems to have been a lucky coincidence. Four other Chinese earthquake predictions also worked. They, too, were flukes. And back in 1994, a student of mine claimed to have predicted the Northridge earthquake in the San Fernando Valley region of Los Angeles forty-eight hours before it happened. He had an aviary in the middle of his house and he claimed that the pheasants were trying to tell him something. He and his housemates left the area. Their house was demolished. Most other predictions made since have been false, and large earthquakes have occurred unexpectedly. Two examples: (1) the New Madrid earthquake was falsely predicted to occur on December 3, 1990; and (2) the shallow quake of magnitude 6.0 that struck northern Italy just north of Bologna in May 2012 was totally unexpected. With all the geoscience advances of the last hundred years, we cannot predict individual earthquakes reliably and accurately. We know where they will happen, but not when. There have been some amazing predictions that have saved thousands of lives, but they were flukes all the same.

Charles Richter wrote in the *Bulletin of the Seismological Society of America* (1977), "I have had a horror of predictions and predictors. Journalists and the general public rush to any suggestion of earthquake prediction like hogs toward a full trough. . . . [Prediction] provides a happy hunting ground for amateurs, cranks, and outright publicity-seeking fakers."[9]

We cannot anticipate all harmful coincidences, yet with or without warnings, we can assess the risk that the worst might happen.

Chapter 14

Psychic Power

How does one mind's electrochemical signals influence another's?

In *Why People Believe Weird Things*, Michael Shermer tells of his visit to an organization called the Association for Research Enlightenment (ARE) in Virginia Beach, Virginia. The organization is a school and repository for the work of Edgar Cayce, a prominent twentieth-century psychic, a school that has been teaching psychic powers since 1931. Attending a lecture on extrasensory perception (ESP) and psychic powers, Shermer volunteered to be a receiver of psychic messages. The instructor explained to his students that some people are born with psychic powers, and others just need practice.[1] After being given a score sheet on which to write down the results of received messages, Shermer and the thirty-four other students were told to concentrate on the sender's forehead. There were two trials with twenty-five messages for each. Each message was one of the following five possible symbols: ✚ ☐ ☆ ○ ≋ . On the first set, Shermer honestly tried to receive and record the message, but on the second set, he just marked all messages as the ✚ symbol. His score was 7 on the first set and 3 on the second.

According to ARE, a score of anything above 7 indicates that the receiver has ESP. First, to have the experiment lean slightly away from the absurd, there should have been a sixth symbol—blank, for the person who just does not get a message. Second, with the blank agreed, we could perform an experiment to help us understand the chances of pairing six symbols: paint two cubes with six symbols on each cube. Every time a message is sent, have a student roll both cubes, and mark whether both cubes land on the same symbol.

The probability of both cubes landing on the same symbol is 1/6, because there are thirty-six possible outcomes and only six possible doubles. What happens when each of thirty-four students rolls the pair of cubes twenty-five times? And, in a group of thirty-four student receivers, how often should we expect to see doubles happening seven times? Ah, we begin to see that there is a bell-shaped curve here, indicating that there is a pretty good chance that random students will be correct seven times. In other words, if you were to randomly select the message symbols, you would have a pretty good chance of getting between three and seven correct in twenty-five tries. It turns out that anyone has a better than even chance of getting more than five correct.

It may seem that communication of just five symbols does not amount to any serious communication. After all, almost any single sentence in this chapter is far more complex than the signals that could be represented by just five arbitrary symbols. But thinking that way would miss the point. If ESP really works with just these five symbols, then it should be considered communication. Hearing ten decibels of the musical notes G and E played on a piano is not the same as hearing the four-note opening motif of Beethoven's Fifth, but it is hearing. After all, Alexander Graham Bell's first successful telephone experiment conveyed a very simple nine-word message shouted loudly into a mouthpiece: "Mr. Watson—come here—I want to see you." That was on March 10, 1876. The scratchy transmission of speech was barely audible on Thomas Watson's end. Who then would have believed that voice could travel electronically, and who then would have believed that we could have wireless personal phones that could transmit voice from anywhere in the world to anyplace else in the world? So, we must be careful in what we believe and what we do not believe. Perhaps the telepathy of just five symbols is just an indicator of understanding yet to come. It's an old concern: the premature popular prejudices of nature. Elizabeth Gilbert raised them in her novel *The Signature of All Things* when she wrote, "Wallace wrote that the first man who ever saw a flying fish probably thought he was witnessing a miracle—and the first man who ever *described* a flying fish was doubtless called a liar."[2] The Wallace in the novel is British naturalist Alfred Russel Wallace, and the allusion is to a true case of a British marine officer returning to England claiming to have seen flying fish in Barbados. But in in real life, Wallace was the discoverer of the *Rhacophorus nigropalmatus*, the flying frog, found in the tropical rain forests of Malaysia.[3]

Extrasensory perception, an expression that includes telepathy and clairvoyance, is one of those action-at-a-distance theories that involve mind transmission and reception of information through uncommon physical senses. Intuition would be a reasonable interpretation of what it means, yet it also means a way of receiving information from channels peripheral to current scientific knowledge. For some true believers, those channels connect the present to the past and the past to the dead. Despite almost a century of repeated negative results of statistical experiments on the existence of human ESP capabilities, parapsychologists have yet to abandon the idea of human ESP capabilities.[4]

Many of the more famous psychics are well connected to outlets for media attention. Kenny Kingston, "psychic to the stars," hosted a radio talk show and was a regular guest of Merv Griffin and *Entertainment Tonight*. Kingston promoted his psychic hotline through infomercials claiming his connections with such celebrities as John Wayne, the Duke and Duchess of Windsor, and Marilyn Monroe. He made millions from his $400-a-session contacts with the dead that included Errol Flynn and

Orson Welles, who were still to be found at the Musso & Frank Grill, the Hollywood restaurant that had been around since Flynn frequented it when he was still alive. You will not catch me saying that Kingston is a fraud; he may be, or he may not be. Wouldn't it be nice if mediums could perform séances to talk with the dead, and predict the future?

Once upon a time, not so long ago, people swallowed magnets to attract love. Why not? Since magnets have miraculous action-at-a-distance powers, it's not difficult to understand how people could believe that souls could be attracted by that incomprehensible magnetic force. We, in typical condescension and misunderstanding of outmoded lore, think that to be whacky. But since the early nineteenth century, we have known that electric currents generate magnetic fields and vice versa. So, we should have been thinking all along that mental activity, which, after all, is electrochemical activity, generates magnetic fields around and outside the human head. In today's acceleration of neuroscience exploration, increasingly sophisticated tools of brain imaging are suggesting notions we would have squinted at just a decade ago. We now have MEG (magnetoencephalography) scanner evidence that emotions expressed in human brains do generate magnetic fields outside the head. Although those fields are relatively weak, it's possible that they, along with brainwave activity, convey signals that piggyback on radio waves to transmit signals far from the source. I don't doubt that that is possible. It may very well be that one person can communicate some signal of love outside his or her brain. Like cell phone signals, those signals could go far. The problem rests with our interpretation of the transmitted signals. Can they be decoded to communicate information to another person? To actually pass on the emotion of love, those signals would have to be decoded to mean not just "love," but "I love *you*" to the receiver. Think of how difficult it is to know of a person's love. If conveying love were just a matter of telepathy of brain signals, every romance novel would be a bore.

Telepathy is the ability to transfer information through some anomalous process of energy transfer unexplained by known physical or biological mechanisms. Such information could be about the past, the present, the future, or contact with the dead. The transference could be emotional kinesthetic sensations through altered states, or it could be through access to the subconscious collective wisdom of the species for the purpose of gaining some intelligence.[5]

Brazil is a country in which 90 percent of its people believe in life after death and in the possibility that the living can communicate with the dead. There is the true story of João Rosa, a crime boss of the small city of Uberaba near São Paulo, and his lover, Lenira de Oliveira. Although Rosa had been seeing other women, he could not accept Oliveira seeing other men. Overcome with jealousy, he followed her and her other man. In the ensuing confrontation Rosa was killed. Oliveira and her new boyfriend were charged with murder. In grief and still in love with Rosa, Oliveira consulted a medium who channeled a letter addressed to her from beyond the grave. At the trial, the defense attorney told the court, "In the letter, channeled by this medium, the deceased confesses. He says his jealousy was the reason for his death. The letter includes details that only people close to him could have known."

Letters from the dead, written by mediums, are accepted by the Brazilian court system as part of the process of delivering evidence. In this Brazilian spiritual climate, no money changes hands. It's all about true belief. The mediums do it for their unflinching belief. In a society that has such staunch belief in the afterlife, the jury is positively receptive to a letter written after death. Of course, Oliveira and her boyfriend were acquitted.[6]

Supporters of the existence of ESP exhibit a few classical cases. There is a famous experiment documented by Upton Sinclair in his book *Mental Radio*. Sinclair believed that his second wife, Mary Craig Kimbrough, had the gift of psychic powers. To test those powers he asked Mary to duplicate 290 pictures as he drew them. Astoundingly, she duplicated 65 successfully, 155 with partial success, and had just 70 failures.[7] But that's just it! You have to count the failures against the successes.

Another famous experiment goes back to 1937. Two people, author Harold Sherman and explorer Hubert Wilkins, telepathized mental images and thoughts by first drawing and writing them in diaries. These telepathies continued daily for 161 days while Sherman was in New York and Wilkins was on an Arctic expedition.[8] On February 21, 1938, they both wrote that cold weather had delayed their jobs, that they saw someone's skin peel from a finger, that they drank alcohol and smoked cigars with friends, and that both had toothaches.[9] Indeed, the two diaries were about 75 percent in common.[10]

The early twentieth century had dignified supporters of ESP, some believing in the psychic powers for reaching the dead. We mentioned Sinclair and Wallace, but imagine the dominant influence of such distinguished people as William James, Henri Bergson, Sir Arthur Conan Doyle, Aldous Huxley, Jules Romains, H. G. Wells, Gilbert Murray, Arthur Koestler, and even, to a certain extent, Sigmund Freud. These eminent psychologists, philosophers, and writers were able to sway others to uncritically join the band. They were not cranks, but rather sincere men who took their works seriously under twentieth-century standards of scientific convention, but without any backup of critical orthodox experimentation.

By the 1930s universities and journals were taking psychic adventures seriously. Duke University had wooed and recruited psychologist William McDougall from Oxford and Harvard to head up a lab that would perform experiments in search of psychic forces. At least two academic journals published articles in support of animal clairvoyance, a telepathic cat, and a mare that could spell out telepathic messages by touching her nose to lettered and numbered blocks.[11]

Husband-and-wife authors Joseph and Louisa Rhine wrote about a horse study in the *Journal of Abnormal and Social Psychology*: "Nothing was discovered that failed to accord with [telepathy], and no other hypothesis proposed seems tenable in view of the results."[12] Perhaps inspired by Arthur Conan Doyle's lectures on telepathy, the Rhines were following Sherlock Holmes's *The Sign of the Four* dictum: "Eliminate all other factors and the one that remains must be the truth." In truth, it does come down to eliminating *all* other factors. The hard part is knowing when there are no factors left to be eliminated.

I am reminded of a preposterous statement in David Auburn's play *Proof*, which had a popular following a few years ago, where Hal, a mathematician studying the proof of a theorem, says that he cannot find anything wrong with the proof, so it checks out. This is logically equivalent to saying that if it is not true, then he can find something wrong. Lewis Carroll's Cheshire cat might agree with a grin. He's the one who said that dogs are not mad and that he is not a dog, to conclude that he is mad. That kind of logic can happen only in Wonderland.

At the heart of ESP is what parapsychologists call the *psi* phenomenon. Psi is the twenty-third letter of the Greek alphabet, though presumably launched as a phonemic resembling the first syllable of *psyche* to represent mental interactions that cannot be explained by established physical principles. Twentieth-century philosopher of science Charles Dunbar Broad argued that the existence of psi events is in conflict with science laws at the fundamental levels of space, time, and causality. His 1949 paper in the journal *Philosophy* offers nine points by which psi is in conflict with conventional reasoning and with the physical laws as we know them.[13] Psi supporters agree among themselves that such phenomena are completely incompatible with modern physics, and yet they accept such a paradoxical conflict. Rhine argued, "Nothing in all the history of human thought—heliocentrism, evolution, relativity—has been more truly revolutionary or radically contradictory to contemporary thought than the results of the investigation of precognitive psi."[14]

In 1937 Ronald Aylmer Fisher wrote a book on the design of scientific experiments with rigorous numerical measures for distinguishing flukes from outcomes that could lead to reliable predictions.[15] His purpose had nothing to do with refuting ESP. Rather, it was to teach, in very elementary terms, the idea of how we should accept or reject coincidences using raw data.

Fisher gave a fictional account of an English tea party where a lady was overheard saying that she could tell by taste whether milk had been added to her cup before the tea or afterward. No doubt, such a claim would take a finely discriminating palate. Fisher's imaginative account led him to design a possible experiment. In the real world we might ordinarily hold the woman precisely at her word, but in a more reasonable mathematical model we would be more inclined to be more flexible and suggest that *most often* she could distinguish whether milk had been added before or after tea. Fisher understood that even events that happen *most often* could happen by purely random circumstances. He really meant his essay to be about the design of experiments and the concern over subjective error, but he was also aiming at the connection between ideal mathematics and imperfect real world experiments.

The experiment involved eight cups of tea, four with milk added before the tea and four afterward. Clearly, if she were right with all eight cups, the experimenters would be convinced that she could discriminate. But what if she missed one? Would that contradict her word? Maybe not, but what if she missed two?

Mathematics can be used to determine the outcome. The lady, in the exuberance of her extraordinary claim, should have permitted herself some possibility of error. (Wouldn't the world be wonderful if we could all do that from time to time?) After all, her taste buds would have changed after the first few sips, and so would the milk molecules. With such a delicate difference between tea poured before milk and tea poured after milk, it seems only fair to relax her strict claim and permit a few errors.[16]

Modern statistics began in the late nineteenth century. Its premise is that random variables distribute themselves over a range of possibilities. The lady who claims to be able to distinguish whether the milk had been added to her cup before the tea or afterward is different from the clairvoyant who claims that he or she could foretell the sex of an unborn child. The truths of their claims come down to distinguishing between random guesses and true clairvoyance. After all, the sex of the unborn child is determined randomly, and so is the guess. The lady who tells us that she can distinguish her teas is directly using physical taste buds along with a confidence in her ability to perceive differences in tastes.

We see coincidences as events that are mysteriously fated by some deeply significant design. We suspect a correlation between two complex phenomena. The real problem is that we naturally tend to make connections were there are none.

This is the stuff of probability and statistics. We do err, and statistics allows a certain amount of flexibility of truth. Statistical approaches are very delicate. According to Fisher, statistical corroboration is evidence of suspected truth. He wrote,[17]

> In considering the appropriateness of any proposed experimental design, it is always needful to forecast all possible results of the experiment, and to have decided without ambiguity what interpretation shall be placed upon each one of them. Further, we must know by what argument this interpretation is to be sustained.

If a supernatural phenomenon such as a psychic phenomenon were to have statistical confirmation, it could then be a good candidate for rational inquiry. But the only statistical confirmation of psi seen so far has come from findings that are heavily dependent on clerical errors, inadvertent sensory inklings, and hyperchance conditions. Until we see legitimate statistical confirmation, psi should be consigned to the world of magic where scientists are comfortable with coincidence and the accepted physical tools of the magician. Although magicians can give their audiences mind-baffling performances that appear to conflict with the known laws of physics—levitating bodies, or piercing them with sharp sabers, or knowing from a distance which card is half-way down a pack—we know that they are tricks of trust, vision, awareness, and reliance on gullibility.

We are asked not to question how telepathized information gets from one brain to another. If science has a say, it would ask for an account of how electrochemical activity in the brain converts to raw data signals capable of traveling through space, and how those signals get reconverted back into electrochemical changes in neurons. American population geneticist George Price mockingly put it this way when describing how a psi phenomenon could transmit information about a certain card buried in a deck: "There is no plausible way to explain these details except in terms of special intelligent agents—spirits or poltergeists or whatever one wishes to call them. The proper target card is selected by a spirit. A spirit implants information in the brain in proper electrochemical form. The ability disappears when the spirit tires of working with a particular person. In short, parapsychology, although well camouflaged with some of the paraphernalia of science, still bears in abundance the markings of magic."[18]

Whenever we are asked not to question truth, we are being asked to accept magic, miracle, or supernatural as the answer. Aside from the tricks performed by entertaining magicians, the word *magic* refers to the notion that coincidences come from supernatural powers, influences that defy established physical laws. A man on a stage transforms a scarf into a white rabbit. Houdini's tricks defied all sensibilities of physical law, yet he scorned the notion of ESP.[19]

The Normalcy of Action at a Distance

The sixteenth century labored to articulate universal laws from Aristotle's physics maxim that everything in the universe had a natural place to which it would strive to return when moved. Before Sir Isaac Newton conceived of the law of universal gravitation, man's fate was somehow linked to the movements of the heavens. From Newton we learned that apples fall for the same reason planets attract each other. Man's fate and the movements of the stars were no longer linked. When Newton was born the first edition of the King James Bible claimed that "The sun also riseth, and the sun goeth down, and hasteth to his place where he arose. The wind goeth toward the south, and turneth about unto the north; it whirleth about continually and the wind turneth again according to his circuits. All the rivers run into the sea; yet the sea is not full; unto the place from whence the rivers come, thither they return again."[20]

In John Milton's *Paradise Lost*, God sends the archangel Raphael down to Paradise to admonish Adam and also to uncover the identity of Satan. Raphael is entertained at a table "with pleasant liquors," the finest fruits and meats of Paradise brought by Eve while Adam asks about the world, how it came to be and how the planets move. Raphael explains,[21]

> *. . . Heaven*
> *Is as the Book of God before thee set,*
> *Wherein to read his wondrous works, and learn*
> *His season, hours, or days, or months, or years,*
> *This to attain, whether Heaven move or Earth, . . .*
>
> *Hereafter, when they come to model Heaven*
> *And calculate the stars, how they will wield*
> *The mighty frame, how gird the sphere*
> *With centric and eccentric scribbled o'er,*
> *Cycle and epicycle, orb in orb. . . .*

Milton completed *Paradise Lost* just before the Great Plague hit London in 1665, when Newton left Cambridge and took refuge at his childhood home in the hamlet of Woolsthorpe, where he discovered, among other things, his universal law of gravitation, the description of the composition of the action of gravitational force with inertial motion that both holds the planets in their orbits and causes the apple to fall.

But by the late eighteenth century, gravity was beginning to be thought of as a possession of systems of matter: two objects attract because they were a certain distance apart and they contained a certain amount of matter. Their attraction was by virtue of their "bulk." Newton thought of gravitational forces as phenomena dependent on their relations with other bodies. A body in isolation has no intrinsic gravitational force, but when another body comes near, it exerts force on that body, and that body exerts a force back.

The prevailing scientific view was that law determines the universe; yet, unlike the motion of the planets, the governing laws of biology are dependent on far too many variables to be perfectly explained. An apple may fall from a tree and abide by Newton's simple laws of motion, while the apple itself is an extremely complex bundle of molecules held together by a formidable number of complicated internal atomic pulls.

We now live in a century where action at a distance is normal. The last century saw the development of radio and television, where sound and picture signals miraculously travel through mostly empty space for thousands of miles piggybacked on radio waves. We have grown accustomed to cell phones and Wi-Fi without questioning how or where the information comes and goes. We no longer question new forms of action at a distance that bring pictures and sound from Beijing to New York in the blink of an eye. For a simple impression of how all that happens, think of how the voice of one person is heard by another.

Figure 14.1. Model of sympathetic frequencies.

There is a wonderful model of how the ear works that mathematician Sir Christopher Zeeman once showed me. Tautly tie a string across a large room. (See Figure 14.1.) On one side, tie several strings of unequal length onto the taut string. At the ends of each hanging string secure a weight of, say, a few ounces. At the other end of the taut string, hang duplicate copies of the hanging weights in no particular order. When the whole system settles down from any motion, carefully pull any weight to the side and let it go. What happens? Except for very minor movements of the system, only two hanging weights will be moving with any significant recognition: the two hanging weights of the same length. Why? Because the frequency of the displaced weight conveys its frequency to the taught string so that any (but only) hanging weights of sympathetic frequencies will resonate.

There is nothing new in this little experiment. Piano tuners use this principle every day to tune the keys of one octave by hitting the keys of neighboring octaves. The overtones of any one note come from the frequency vibrations of piano strings with sympathetic frequencies.

And that is precisely how the human ear works. Russian mezzo-soprano Olga Borodina sings the aria of Dido's Lament from *Dido and Aeneas*: "When I am laid in earth, . . . " She projects notes from her larynx that cause waves of air in front of her mouth. Those waves move through space until they reach the ear of a person in her audience. In the cochlea of that person's ear, the cilia that are partially immersed in fluid sympathetically move in resonance with the wave of air. The moving cilia create a fluid motion that converts into electrical signals, which in turn excite auditory nerves.

People in ancient times must have pondered over how the voice of one person could be heard by another over a space with no apparent mechanical connection. As a child whose comic book hero was Dick Tracy, I marveled with some skepticism over where in the world my hero got his wristwatch picture telephone. These days, Dick Tracy's watch is yesterday's technology—it could only get picture telephone, big deal. We don't even begin to pay attention to how our cell phone signals get across empty space, or how our e-mail messages get from one side of the planet to the other in just a few seconds.

Mr. Wonka, in Roald Dahl's *Charlie and the Chocolate Factory*, was not fazed by the phenomenon when showing his marvelous invention to Mike Teavee.

> "Now then!" he said. "The very first time I saw ordinary television working, I was struck by a tremendous idea. 'Look here!' I shouted, 'If these people can break up a photograph into millions of pieces and send the pieces whizzing through the air and then put them together again at the other end, why can't I do the same thing with a bar of chocolate? Why can't I send a bar of chocolate whizzing through the air in tiny pieces and then put the pieces together at the other end, all ready to be eaten?'"[22]

Imaginably, Mr. Wonka is far ahead of the game in understanding action at a distance, and possibly ahead on the theory of everything.

Coincidence Without a Cause

Action at a distance is at the core of extrasensory perception. I would not be surprised to find out that humans *do* possess some means of having perceptions beyond the usual five. Some people are extremely sensitive to atmospheric pressure, and some have keen detectors for social cues. Possibly some people have a relatively solid sensitivity to radio waves. I would not doubt that. But there is a long road from that sensitivity to the ability to code and ethereally transmit messages from one mind to another.

Assuming we don't abuse the planet to the point of self-destruction, we are in the infancy of human existence. Believing otherwise would be smug and unwise. We must assume also that we are in the infancy of what we understand about physics and nature. We have theories for many things, but it will be a long time—perhaps millennia, perhaps never—before we spot the boundaries of the theory of everything. And yet the resolution of scientific discoveries is always improving.

Chapter 15

Sir Gawain and the Green Knight

IN REAL LIFE, any happenstance with extraordinarily low probability might seem like a once in a lifetime event, and yet people do win lotteries two, three, or even four times in a lifetime. In folklore, legend, and fiction, far more extraordinary events with far worse odds frequently happen. Stories often defy the odds because the storyteller in charge is always ready to bring us to our toes by suspension of belief.

Flukes and coincidences often blur the distinction between fact and fiction. In folklore, legend, and literature we tend to suspend belief in reason so as to enter a world that is not our own, an illusory world where we are the ghostly observers of events that tell us something about ourselves as humans. Like most fictional accounts, the stories here, with their embedded flukes and coincidences, show us who we really are as archetypes in the big picture.

"A certain man once lost a diamond cuff-link in the wide blue sea," Vladimir Nabokov wrote in his novel *Laughter in the Dark*, "and twenty years later, on the exact day, a Friday apparently, he was eating a large fish—but there was no diamond inside. That's what I like about coincidence."[1] The passage is characteristic of Nabokov's delightful wit. It's not a long paragraph, and yet, as we read it, we find our thoughts quickly building to anticipate something that does not happen. Nabokov set us up with an expectation, hits us with a surprise, and ends with, "That's what I like about coincidence." It is fiction! In fiction, anything can happen.

The passage tells us what a coincidence really is. A surprise. Only in this case, the surprise is that there is no surprise. Surprise is a fundamental structural element of storytelling, and coincidences, by definition, always carry surprises. Anthropologists tell us that ever since humans built the minimum of language sophistication to tell a story, they told stories. Every culture on earth has told stories to its children. Those stories may have some buried truth originating in reality, but it is the depths of imagination that makes them live. Stories of legendary heroes particularly use coincidences for character meetings.

Many years ago when I was a student in Paris, I lived for one week in the Hotel Albe on the corner of two very narrow streets, rue de la Huchette and rue de la Harpe. These days the Albe is a four-star hotel, but back then it was a squalid place with a broken one-person elevator, miniature rooms, saggy mattresses, and tepid water in shared baths. The neighborhood was just the place for a student with little money and few friends. Just a few meters down the street was Théâtre de la Huchette, a small theater showing Eugène Ionesco's play *La Cantatrice Chauve* (The Baldheaded Singer), whose English title became *The Bald Soprano*. I walked further down the street to Shakespeare and Company to find a copy of the play in English. Reading it and seeing it several times for a franc was my way of learning French better than my raw get-by grasp of the language.

By my count there are thirteen illusory coincidences in the play. Elizabeth and Donald Martin are at a dinner. They don't seem to know each other, but think that they have met someplace before. Donald asks whether they may have met by chance in Manchester. He left Manchester just five weeks before, on the 8:30 morning train. So had Elizabeth.

This dialogue continues through an array of the Martins' phantasmagoric coincidences. In the end the Martins find out they both live on the same floor of the same flat, and indeed in the same bedroom. They sleep in the same bed. Elizabeth is stunned! She says it is possible they met the previous night in Donald's bed, though she does not recall it. Donald then tells her that he has a blond two-year-old daughter by the name of Alice who lives with him. She is pretty and has one white eye and one red eye. To this Elizabeth responds with astonishment that it is quite a coincidence, for she, too, has a pretty two-year-old named Alice who has one white eye and one red eye.[2] Clearly, this is the theater of the absurd and these coincidences are surely absurd without some clinical hints of dementia.

Coincidence in fiction is not the same as coincidence in real life. In fiction, the author is the cause. Sometimes, generally in bad novels and in excellent ones also, coincidences happen without the direct intention of the author—a chance meeting that springs into the plot. Intended or not, they arouse cognitive effects that might otherwise lead to variable paths of understanding.[3]

Legends

The timeless poem *Sir Gawain and the Green Knight* comes to us from a late fourteenth-century parchment codex now in the British Library. It's a romance, a skillfully told fairytale, a story of fidelity and courtesy, a dark story of the underworld, and a true wonder. The author himself correctly tells us that it is "an outrageous adventure of Arthurian wonders."[4] It is told through a mesh of circumstances and at least one clearly astounding coincidence.

It begins on New Year's Eve. That already is a coincidence, since the Green Knight, like the year itself, is seemingly about to die and come back to life. A party has been going on for fifteen days and nights. But on this New Year's Eve that awesome fellow "who in height outstripped all earthly men," the Green Knight, carrying a green battle-ax on his green horse, rode directly into the party at King Arthur's court.

As the sound of the music ceased,
And the first course had been fitly served,
There came in at the hall door one terrible to behold,
Of stature greater than any on earth;
From neck to loin so strong and thickly made,
And with limbs so long and so great
That he seemed even as a giant.
And yet he was but a man,
Only the mightiest that might mount a steed;
Broad of chest and shoulders and slender of waist,
And all his features of like fashion;
But men marvelled much at his colour,
For he rode even as a knight,
Yet was green all over.[5]

In an outrageous challenge to the Knights of the Round Table, the Green Knight dared anyone to cut off his head in a single stroke of his green ax. Then came his caveat: whoever was to succeed would have to appear at the Green Chapel (a three-day journey from the court) on the next New Year's Eve, when the victor would offer himself for decapitation. A peculiar dark story, indeed!

In case you don't know the story, I will not give away the surprise ending. Sir Gawain, a knight of the Round Table, decapitates the Green Knight with a single stroke of the mighty ax. Did you think he couldn't? The Green Knight's head drops to the floor and rolls a bit with blood dripping from it. But the knight's body, blood spurting from the wound, calmly picks his head up by its hair, takes up his bloodied weapon, and mounts his great horse, and, through his head's moving lips, reminds Gawain about the other end of the challenge—

> *"Look, Gawain, that thou art ready to go as thou hast promised,*
> *And seek leally till thou find me,*
> *Even as thou hast sworn in this hall in the hearing of these knights.*
> *Come thou, I charge thee, to the Green Chapel,*
> *Such a stroke as thou hast dealt thou hast deserved,*
> *And it shall be promptly paid thee on New Year's morn. . . ."*[6]

And so, a few days before the following Christmas, Sir Gawain goes off in search of the Green Chapel. At this point we get to the magic of the tale. You would think that Gawain had enough time to find out more about this Green Chapel, or at least where it is. But no! He mounts his steed Gringolet, and magically heads for Wales without a clue as to where the Green Chapel is. He asks everyone he meets along the way, but no one knows.

> *And ever he asked, as he fared, of all whom he met,*
> *If they had heard any tidings of a Green Knight*
> *In the country thereabout, or of a Green Chapel?*
> *And all answered him, Nay,*
> *Never in their lives had they seen any man of such a hue.*
> *And the knight wended his way by many a strange road and many a rugged path,*
> *And the fashion of his countenance changed full often ere he saw the Green Chapel.*[7]

And now comes the elemental coincidence. It's Christmas Eve and Sir Gawain finds himself lost in a great forest. He prays to the Virgin to show him a place to take shelter, and magically (although the Gawain poet might say, "guided by God") stumbles on a great castle. A lord of "stupendous size" and the lady of the castle courteously greet him, and make him comfortable. The lady's beauty, Gawain notes, excels Guinevere's. On each of the three dawns before New Year's Day, the lord leaves to go on hunts, and returns at dusk. On the first two mornings the dazzling lady steals to Gawain's bed and talks to him in an irresistible voice. Gawain is immovable and grants her just a kiss on the first day and two on the second—steadfastly, nothing more. What a man! This guy is to have his head cut off the next day. Who among us could be so spotless?

On New Year's Eve morning she insists that Gawain should accept a heavy ring as a gift. But he knows that receiving such a gift would commit him to be her knight, to surrender his being, and to resign his chivalric commitments. He does not accept. She offers him a token, her girdle of green silk with golden lace. He is about to decline this, but she says, "For the man who binds his body with this belt of green, / As long as he laps it closely about him, / No hero under heaven can hack him to pieces, / For he cannot be killed by any cunning on earth." Now how can he not accept *that* silk?

There is plenty more to the story, but in the end all of his trials were part of the game. And in the end we find that the lord of the castle is the Green Knight. The ax is lifted and put down twice. On the third lift, it comes down and grazes Gawain's neck, hardly leaving a scratch.

What do we make of all this? The Green Chapel is but 2 miles from the castle. Gawain travelled about 36 miles to the castle, possibly.[8] Why 36 miles? The poem mentions that Gawain was on his way to northern Wales. Legendary Camelot was anywhere in Britain. But celebrated Arthurian scholar William Raymond Johnston Barron claimed that, in this particular poem, Gawain started off from the Cheshire/Staffordshire border. That being the case, my Google Maps tell me that the shortest distance traveled would be about 36 miles. How fortunate that without knowing where he was going when he starts out from Camelot, other than heading toward northern Wales, he accidentally finds himself within 2 miles of his target.

It is a colossal coincidence. Just imagine trying to do that yourself. But it is a fabricated coincidence that writers often employ to develop a plot where the mood of peculiar circumstances move along arcs of reasonable logic. This is almost typical of legend writing, and somewhat necessary. The poet, whoever he is, was forced to have Gawain lost in the great woods and accidentally (or divinely) stumble upon the great castle. If he knew the way, he would know the castle. And if he knew the castle, chances are he would know the identity of its lord. The power of the story is built from Gawain's not having that information. Forgive me if I've just given the end away. It is a very old story, but a Western one. Eastern stories play the game differently. Eastern folklore is filled with stories of coincidence that are perceived as magical events. They are the stories of Hindi gurus, Tibetan monks, and other solons of a more universal literary culture.

Western folklore has parallels, but often with religious overtones where magic is viewed as miracle. The borderline between folklore and religion in Western culture is misty, with religious stories designed to demonstrate the power of God's will. They are the stories of Judaic-Christian sages, Greek oracles, and prophets of major religions. Greek oracles, for instance, tell coincidence stories coming from reasonably trustworthy historical writings and Greek lore. The writings of Plutarch, Xenophon, and Diodorus that talk about oracles are taken as effectively true. Interestingly, almost all documented oracles have correctly predicted futures by coincidences in favor of those predictions. Of course, as with any successful fortuneteller, those prophecies were worded ambiguously in order to convince believers that the oracle possesses legitimate power.

Folklore is a proto-psychological accounting of the human need to pay attention to surroundings, to what is strange and to what is not. It speaks to one of the primal urges that helped our primitive ancestors survive the terrors of the wild. The recognition and underscoring of coincidences gives warning to the tribe that anything can happen. It embellishes the legend, brings us face-to-face with believable events where good and bad accidents really do happen, and adds a sensitivity of unprotected risk to the folklore hero's daily ventures into the unknown.

The folklore of healing filters through an imaginary line dividing tale and real life. Physical ailments—blindness, lameness, and hunched back—are magically cured by fictional design to demonstrate the powers of gods or wizards as well as the power and control of those who see themselves as the conveyers of some superhuman will. Science, logic, and reason are sidestepped in favor of fate, explainable only through sequences of coincidences. Folklore gives us a heightened awareness of those possibilities of coincidence. Consider the Chinese folklore belief known as the Red Strings of Fate: every child born has an invisible (to humans) red string attached to an ankle, with the other end similarly attached to an ankle of the child's intended spouse. The matchmaker god decides the fate, and it is he who ties the string that stretches and tangles but never breaks. It is the Eastern version of a fated life: a long series of coincidences that must happen if one person is to find his or her predestined mate. There was a time when the Red Strings of Fate had a ring of truth about it. That would have been a time when people would not stray far from their home villages, a time when those people had close connections for most of their lives. The string was tied metaphorically as an agreement between sets of parents. That metaphor did not have the power of coincidence that it has today when those strings of fate are so prohibitively long and tangled.

The Three Princes of Serendip is often referred to as an example of serendipity. In fact, the very definition of our modern English word *serendipity* comes from the title of that fairytale. Originally published in Venice, it was translated from Persian and Urdu to Italian in 1557. It comes from "The Eight Paradises" written by Amir Khusro (a.k.a. Khusrau) of Delhi back in the early fourteenth century. The story itself may be older still, and likely based on the life of Bahram V, a Persian king of the fifth century. We know this story from the Fourth Earl of Oxford, a man by the name of Horace Walpole, who happened to be an antiquarian and a famous author of his time. According to Richard Boyle, an expert on the British colonial period in Sri Lanka (in those days called Ceylon), and contributor to the *Oxford English Dictionary*, it was Walpole who claimed he came across "a silly fairy tale called *The Three Princes of Serendip*."[9] It's a story that was known in Europe since the late twelfth century. There are many versions of the tale, the so-called riddle poems—*The King and the Three Brothers*, *The Inheritance of the Three Sons*, *The Clever Bedouin Reads Footsteps in the Sand*, *Three Clever Brothers Before the Judge*, *King Solomon and the Three Brothers*, and *King Solomon and the Three Golden Balls*.[10] It's a story of three brothers who wander through the countryside accidentally encountering riddles that they solve in whimsically clever ways. As we shall find out in the story, those accidents are more coincidences than flukes. Again, according to Boyle, in a letter to Horace Mann dated January 28, 1754, Walpole wrote, "[The brothers] were always making discoveries, by accident and sagacity, of things they were not in quest of."[11] Hence the *Oxford English Dictionary* has as its entry for the noun *serendipity*:

> The occurrence and development of events by chance in a happy or beneficial way: "a fortunate stroke of serendipity."

The three princes could have been the sons of Bahram V or the sons of Giaffer. And Serendip (or, as sometimes spelled, Sarendip) is the ancient name for Sri Lanka.[12]

The story begins with these words:[13]

> In ancient times there existed in the country of Serendippo, in the Far East, a great and powerful king by the name of Giaffer. He had three sons who were very dear to him. And being a good father and very concerned about their education, he decided that he had to leave them endowed not only with great power, but also with all kinds of virtues of which princes are particularly in need.[14]

So Giaffer banishes his sons from the kingdom of Serendip so that they may gain some worldly wisdom on top of their book learning. They come to the kingdom of the great and powerful Beramo. There they have many adventures and make many discoveries though observations and inferences. The first incident is an encounter with a camel driver who stops them on the road to ask whether they have seen his missing camel. (In Europe, the story is about a mule; in India it is about an elephant.) They had not. But in showing off intelligence they ask the camel driver three questions. Was the camel blind in the right eye? Was it missing a tooth? Did it have one lame leg? Yes, the camel did have all those ailments. So the princes tell the driver that they had seen such a camel on the road. The driver rushes down the road to fetch his camel. Without success, he once again encounters the three princes, who tell him the camel was laden with butter on one side and honey on the other, and ridden by a pregnant woman. At this point the driver becomes suspicious that the princes have stolen his camel. It is a silly story that requires us to guess why the driver should be suspicious. We can only speculate that since the princes knew so much about the camel they must have seen it, and since there was no sign of the camel anywhere, the princes must have stolen it.

The driver has the princes brought before a magistrate. They swear that they'd never really seen the camel. When the magistrate questions how they could have possibly known so much about the camel, they confess to having observed clues (ones they had not been looking for) to infer essential details that just happened to be coincidental with the facts. In the end the camel is found and the princes are asked to divulge how they inferred the unusual characteristics of the camel.

The explanations are all quite silly. The camel was blind in its right eye because the grass on the left was eaten, not the grass on the right. It was missing a tooth because each bit of grass left a small wisp of grass with each bite taken. The camel was laden with butter on one side and honey on the other because flies occupied one side of the road and bees the other. Prints in the road showed one dragging foot. And what about the pregnant woman? The princes claimed to have felt carnal desires as they passed the spot where they saw the woman's footprints. Carnal desires? All silly reasons. The point here is that, from the beginning, the princes were walking along a road observing things that became relevant only after the incident of meeting the camel driver. In other words, they were accidentally observing things that they could not have anticipated a use for. They were not in search of a lost camel before the camel driver told them that his camel was missing.[15]

Yes, it is an example of serendipity, but it is also an example of coincidence, an exotic, entertaining tale. What brought together such keen observations so long before meeting a camel driver? One answer might be that they were incredibly observant of their surroundings, naturally paying attention to everything they passed—grass, flies, ants, and marks on the road—all in anticipation of needing that information for a later time. But another answer is that they had made a wild guess that was backed by the intelligence of observation. There could have been any number of reasons for the grass to be eaten in clumps on one side of the road where flies habituated. The fact that the camel driver's missing camel had all the characteristics described by the three princes seems more likely a coincidence supported by some intelligence, and keen observation unintentionally committed to memory.

Meaning in Fictional Coincidences

John Pier and José Angel Garcia give the following definition of coincidence in their book *Theorizing Narrativity*:[16]

> "Coincidence" is equally related to happenings, namely as the unforeseeable and (seemingly) inexplicable yet apparently meaningful intersection of two occurrences, sometimes even of two causal chains or sequences of happenings and events previously introduced into the story-world, but without causal connection to one another.

This definition permits causal chains, not necessarily direct causal connections. But an unexpected chain of events where the causes are lost at places in the chain gives a heightened surprise that makes any resulting coincidence seem real. The definition also explicitly requires fictional coincidences to be meaningful, as they more or less always are.

Fictional characters are often intersecting in space and time with no apparent cause, through circumstances that are necessary to make the story plot make sense. Those characters might have had some relationship before their odd circumstantial time and space narrative intersection. That old relationship needn't have been a physical meeting. It might have been an old fling, a kinship, some hostility, or just a collegial acquaintance. The coincidental "meeting" would signify little, if it had not had meaning through recognition of each character's importance to the plot. Any link between the previous relationship and the physical meeting should appear to be unconnected without any appearance of being causal, for otherwise the narration loses the desired effect of witnessing the odd unfamiliar along with the intended cognitive delight of trying to make sense of the coincidence that just happened. Delayed recognition is one tactic. I suspect that when authors use such tactics intentionally, they are hoping to create emotional impacts that place the identity of individual characters into the larger plot.

I also speculate that sometimes authors subconsciously include minor details, events, symbolic metaphors, or scenes that end up having more meaning than the author intended. It can be argued that several subconscious aspects of an author's life are responsible. It can also be argued that, as in real life, we are all connected by those proverbial six degrees of separation so that everyone ends up connected in ways for which they have no sensible explanation. Freud had much to say about this, and so did Jung. There are many examples. Some unintentional details have found their way into my own works. Are they flukes, or words escaping from the subconscious? It could be argued that such subconscious inclusions are not a *surprising concurrence of events with no apparent causal ties*; however, one could also argue that words on a page come from a concurrence of subconscious and conscious elements.

In literature, the conscious track has a lag time. Read Dostoyevsky's *Crime and Punishment* and come to the point when Raskolnikov kills the old woman with a swing of an ax. What role does the ax play as we read further? Why did Dostoyevsky decide that the old woman should be killed by an ax and not by a gun or bludgeoned to death with a poker? The reader's psyche would respond differently had another weapon been used. An ax has connotations very different than a bruising to death. It leaves readers with contradictory emotions and clashing images in the mind: a gruesomely bloody death, and a humanely swift death. In other words, a mental impression of the crime would have been quite different had the old lady been killed in any other way. Or Dostoyevsky's choice could have been just a coincidence of the moment while he was writing the scene. We could ask the same question of the Green Knight. Why a heavy green ax, when a mighty sharp sword would do?

A contemporary example is Paul Aster's *Moon Palace*, a book rich in flukes and phantasmagorical coincidences that happen to the narrator, Marco Stanley Fogg. The coincidences are so improbable, that Marco himself has trouble believing them. After several weeks of living a penniless, malnourished life and sleeping in the bushes of New York's Central Park, a friend finds Marco practically at the point of death. After rehabilitation, Marco answers a job ad typed on an index card posted at the Columbia University student employment office. The job calls for a live-in companion to an old, blind, cantankerous invalid named Thomas Effing. A few months go by before Thomas begins to plan his own obituary and asks Marco to write it. Back in 1916 Thomas's name was Julian Barber, and that's when the obituary story begins, at a time when Julian felt that he had to get away from his mentally troubled wife.

Julian travels to a remote area of Utah. He comes upon a hermit's cave filled with provisions, comfortable furniture, and several loaded riffles. He finds the hermit dead, recently killed from a gunshot wound, and notices that the hermit looked just like himself. So, he buries the hermit and plans a new life with a new identity, and spends the winter months in the cave. In the spring a visitor appears, George Ugly Mouth, a Native American who thinks Julian is his friend, the hermit. George tells Julian that three Gresham brothers, a gang of train robbers, are on their way to the cave, their hideout. Julian suspects that the gang killed the hermit. The gang returns, Julian shoots each of the three brothers and runs off with $20,000 of their stolen money. He returns to civilization under his new name, Thomas Effing, learns that his wife bore a son before he left for Utah. The son, Solomon Barber, grew up to be an American history professor at some small midwestern college. We learn that Solomon always thought that his father died in an accident somewhere in Utah. We also learn that Solomon was dismissed from his job after a scandal about his sleeping with one of his students. The young student disappears, and twelve years later is fatally run over by a bus. After Thomas dies, Marco writes to Solomon to tell him that his father had died and had left him a great deal of money. Solomon meets Marco in New York, and tells Marco that back in the forties he had a student from Chicago by the name of Emily Fogg.

> "One coincidence after another," [Marco] said. "The universe seems to be filled with them."
>
> . . .
>
> "She was a beautiful and intelligent girl, your mother. I remember her well."[17]

In real life one might question the odds. But this is a fictional story, with no confident formula that could give us the probability that Marco's story would center on such a colossal coincidence. There are, though, a few investigative methods of narrowing the playing field. Fiction has advantages that real life does not: a carefully constructed plot and a strategically chosen setting. For the most astonishing coincidence in *Moon Palace* to work, the setting had to be a very large city. There aren't too many options for that. And if New York is the choice, then Columbia University is also the choice. The field narrows significantly to just a neighborhood of New York—roughly, a mile area centered at 116th Street and Broadway, while still leaving open a vast number of interfering directions and possibilities.

In real life, the question would be: how many young people in New York who have never met their father, accidentally came in contact with their father through some fluky encounter, say, last year? If we could have a show of hands from all young folks residing in New York City, we would likely see at least a dozen raised hands. Those people might not have a great memoir to write, but their coincidences could make some fascinating stories. They would tell us that they met their fathers by some wild coincidence. It's a huge city with lots of people, lots of odd connections, and lots of opportunities for synchronicity. New York provides a latticework of chance meetings linking the past, present, and future in ways we can fathom only by understanding both the hugeness of population and the myriad combinations of pathways connecting one person to another.

I suspect that if we questioned acclaimed fiction writers about their choices in constructing particular incidents in their works, they would tell us that some of their scenes were constructed by fortunate coincidences of the moment. But then there is a phenomenon that psychologists call the *priming effect*, which tells us that our actions and our emotions are affected by the experience of recent events. For instance, if you were asked to fill in the blanks of the word S_ _ P, you would very likely write SOAP if you had just washed your hands; you would very likely write SOUP if you had just sat down for dinner. So, it might be that some of our comprehension is carried by the coincidences between the words we read and our most immediate experiences. Life is like that. Our thoughts and actions seem to be primed by chains of experiences, and yet fate has its odd ways of stepping in to tweak and perturb the balance.

Epilogue

WE TEND TO THINK of the world as being small and large. On the one hand, it is no bigger than our neighborhood, our friends, acquaintances, and possibly our limited travels. On the other, it is as vast as it is from the window of a flight over the midlands of England, or the limitless forests of Maine. They give us contradictory impressions of just how our intuition should react to the many possible flukes and coincidences that could happen. We bump into our friends in the vastness of the world as if the whole world were just a small town; we—the family of lottery players of the world—win multiple lotteries because our small world is actually quite gigantic.

The world is mind-bogglingly vast. Its people are tightly clustered, not just in cities, but also in space-time of their connections. So, of course, those seemingly unlikely events happen because of the vast number of experientially available possibilities. Do events coincide merely by chance? Or do we use chance as the excuse for what we do not know? When we look for a cause, we might not see one. But on further investigation and deconstruction, the dots connect.

There are few analyses using serious mathematics to explain the regularity of coincidence, aside from those of Persi Diaconis and Frederick Mosteller. Their theories show that many of the occurrences we think of as strange are merely events that happen in close time spans and large populations. There are a large number of possible events that could happen at any one moment, but there are also a large number of possible simultaneous events. David Hand, a mathematician at Imperial College in London, gave us slightly different, yet complementary, perspective to understand coincidences. His *improbability principle*, a collection of meshing probability laws, each in support of the other, explains why highly improbable events are bound to happen. Most parts of the principle are more qualitative than quantitative, with no actual numerical measures of improbability. Rather, those laws are statistically narrative, giving credence to the idea that improbable things are bound to happen more frequently than we would expect. For instance, the collection contains what Hand calls the law of inevitability, which tells us "if you make a complete list of all possible outcomes, then one of them must occur."[1]

There is just one more coincidence that deserves to be listed, just so I can leave my readers with the quandary of what makes a coincidence a coincidence. Sixty-six million years ago a comet crashed into Earth at high speed somewhere near the Yucatán Peninsula, creating a 110-mile-wide crater.[2] From NASA missions, we now know enough about the composition of comets to know that it was a comet and not (as had been thought earlier) an asteroid. There is an ongoing debate among paleontologists, geologists, and astronomers on what caused the global climate change that killed off the dinosaurs. One theory is that the blast from that comet killed almost all those large lizardlike things we call dinosaurs, along with 70 percent of all other living plants and creatures. Organisms exposed to the intense blast of infrared radiation would have been killed almost instantly. For those species that survived, living conditions, on top of obstructed photosynthesis for plants, over the next 60 million years must have been miserable, an interminably long nuclear winter.

Comets are different than asteroids. They have different chemical compositions, but most significant for our story is that—unlike asteroids—comets travel in orbits. They can go along their periodic trajectories for millions of years without crashing into anything. But when a comet comes close enough to another mass, gravitation perturbs its orbit slightly. It may take another million years for it to return to that other mass for a closer pass. In the case of that notable event sixty-six million years ago, just imagine what might have happened if the orbit of that comet would have been just a thousand yards farther from Earth. A thousand yards on an astronomy scale is minuscule, but enormous when masses are nearby. On the next orbital cycle its mass would have been smaller, so the pull to Earth would have been less. It was this coincidence of orbits that was responsible for a mass extinction of species, and the fortunate birth of our own. It all happened in a matter of minutes, and in a few yards of trajectory difference. And here we are. I leave it to you to judge it a coincidence, a fluke, or divine intervention.

Notes

Introduction

1. A similar definition was first introduced by Thomas Vargish in his *The Providential Aesthetic in Victorian Fiction* (Charlottesville: University of Virginia Press, 1985), 7.

2. *Webster's Third New International Dictionary of the English Language Unabridged*, ed. Philip Babcock Grove (Springfield, MA: G. & C. Merriam Company, 1961).

3. Neil Forsyth, "Wonderful Chains: Dickens and Coincidence," *Modern Philology* 83, no 2, (November 1985): 151–165.

Chapter 1

1. Robert Fiala was professor of media arts at Pratt Institute, a good college friend, and a fine artist. He died unexpectedly in 2009.

2. In Scotland at that time, *stovies* nights were when pubs would offer dishes of free food, usually just fried potatoes, to dodge midnight closing laws. (Restaurants were permitted to remain open past midnight.)

3. Lao-tzu, *Tao Te Ching*, chapter 73, trans. William Scott Wilson (Boston: Shambhala Publications, 2010), 39.

4. Walt Whitman, *Democratic Vistas*, ed. Ed Folsom (Ames, IA: University of Iowa Press, 2010), 67–68.

Chapter 2

1. Charles Dickens, *Bleak House* (London: Wordsworth Classics, 1993), 189.

2. Alexander Woollcott, *While Rome Burns* (New York: Viking Press, 1934), 21–23.

3. On reading Woollcott's telling of the story, the thought had occurred to me that Charles Albert Corliss may have been playing a practical joke on Anne by writing the inscription himself in the few moments when Anne had turned away to look at the towers of Notre-Dame. Woollcott tells us, "There was a moment of silence while her glance drifted along the river to the close-packed green of its islands and the towers beyond. This silence was broken abruptly by his admitting, in strained voice, that after all he was inclined to think that she *had* known the book in her younger days."

4. C. G. Jung, *Synchronicity: An Acausal Connecting Principle* (Princeton, NJ: Princeton University Press, 1960), 22.

5. Ibid., 28.

6. Here is where exaggeration creeps in. Was it really an hour? Or was it just a quarter-hour? These are the typical embellishments that I found with almost all the coincidence stories I investigated.

7. Nicolas Camille Flammarion, *L'Inconnu: The Unknown* (New York: Harper & Row, 1900), 194.

8. Ibid.

9. In itself an impressive work, with hundreds of Flammarion's spectacular engravings, many in color. See https//books.google. com /books?id=ScDVAAAAMAJ&pg=PA163#v=onepage&q&f=false.

10. Nicolas Camille Flammarion, *L'Atmosphère: Météorologie Populaire* (Paris: Hachette, 1888), 510.

11. Flammarion, *L'Inconnu*, 192.

12. Ward Hill Lamon, *Recollections of Abraham Lincoln 1847–1865* (Cambridge, MA: The University Press, 1911), 116–120.

13. My own daughter would sleepwalk when she was young, so I can tell you how frightening it is to witness a real sleepwalker.

14. Gideon Welles and Edgar Thaddeus Welles, *Diary of Gideon Welles*, vol. 2 (Boston: Houghton Mifflin, 1911), 283.

15. Frederick W. Seward, "Recollections of Lincoln's Last Hours," *Leslie's Weekly*, 1909, 10.

16. The calculations for this are complex. The odds of the same person winning a lottery twice have been calculated by Stephan Samuels and George McCabe of Purdue University. They claim that the odds are better than even that some person somewhere in the United States will win twice in seven years. The odds are 1 in 30 that there will be a double winner in a four-month period. I note this here with the understanding that I have not seen the actual calculations. The main source seems to be Persi Diaconis and Frederick Mosteller's paper "Method for Studying Coincidences," *Journal of the American Statistical Association* 84, no. 408 (December 1989): Applications & Case Studies, 853–861.

Chapter 3

1. Arthur Koestler, *The Case of the Midwife Toad* (New York: Vintage, 1971), 13.

2. For this translation of the quote, see Martin Plimmer and Brian King, *Beyond Coincidence: Amazing Stories of Coincidence and the Mystery Behind Them* (New York: St. Martin's Press, 2006), 52–53.

3. Paul Kammerer, *Das Gesetz der Serie* (Berlin: Deutsche Verlag-Anstalt, 1919), 93.

4. Ibid.

5. C. G. Jung, *Synchronicity: An Acausal Connecting Principle* (Princeton, NJ: Princeton University Press, 1960), 105.

6. C. A. Meier, ed., David Roscoe, trans., *Atom and Archetype: The Pauli/Jung Letters, 1932–1958* (Princeton, NJ: Princeton University Press, 2001), xxxviii.

7. Jung, *Synchronicity*, 10.

8. C. R. Card, "The Archetypal View of C. G. Jung and Wolfgang Pauli," *Psychological Perspectives* 24 (Spring–Summer 1991):19–33, and 25 (Fall–Winter 1991): 52–69.

9. David Peat, *Synchronicity: The Bridge Between Matter and Mind* (New York: Bantam 1987), 17–18.

10. Aniela Jaffé, *Memories, Dreams, Reflections* (New York: Vintage Books, 1965).

11. Joseph Cambray, *Synchronicity: Nature and Psyche in an Interconnected Universe* (College Station, TX: Texas A&M University Press, 2009), 12.

Chapter 4

1. Carl Gustav Jung, *Jung on Synchronicity and the Paranormal*, (London: Routledge, 2009) 8.

2. I picked this number because it is the probability of winning the lottery in my home state, Vermont.

Chapter 5

1. These papers remained unpublished for almost a hundred years. See Øystein Ore: *Cardano, the Gambling Scholar* (Princeton, NJ: Princeton University Press, 1953, or New York: Dover, 1965). It should be pointed out that this book of Ore's was the first to expose Cardano's contributions to mathematical probability theory. See Ernest Nagel's review of *Cardano, the Gambling Scholar* in *Scientific American*, June 1953.

2. To extract this in words, it says: the probability P that the difference between the observed probability k/N and the mathematical probability p is less than some small chosen number ε gets closer to 1 as N grows large.

3. G. Galileo (c. 1620), *Sopra la scoperte die dadi* (On a Discovery Concerning Dice), trans. E. H. Thorne, excerpted in *Games, Gods, and Gambling: The Origins and History of Probability and*

Statistical Ideas from the Earliest Times to the Newtonian Era by F. N. David (New York: Hafner, 1962), 192–195.

4. Joseph Mazur, *What's Luck Got to Do with It?: The History, Mathematics, and Psychology of the Gambler's Illusion* (Princeton, NJ: Princeton University Press, 2010), 27.

5. This was first published in 1663.

6. The original letters were edited and published in *Oeuvres de Fermat*, ed. by Tannery and Henry, vol. 2 (Paris: Gauthier-Villars: 1894), 288–314. For the letters in translation, see David Eugene Smith, *A Source Book in Mathematics* (New York: Dover, 1959), 424.

7. Pascal understood that it would be easier to calculate the odds of not throwing a double six. That would be 35/36. He also must have understood that the probability of two independent events happening is the product of the probabilities of the individual events and that, therefore, the probability of not throwing a double six on n throws is $(35/36)^n$. He calculated $(35/36)^{24}$ to be 0.509 and $(35/36)^{25}$ to be 0.494, to conclude that there was a slightly smaller than even chance of getting double sixes on twenty-four rolls of the dice and a slightly better than even chance with twenty-five rolls.

8. $1-(35/36)^{24} < 1/2$, but $1-(35/36)^{25} > 1/2$.

9. That's because the probability of the first die coming up with any one of its six numbers is 1. Certainty. Say it lands on 2. The other four dice must now land on 2. That is a probability of $(1/6)^4$, or 1 in 1,296.

10. See the Numberfile video https://www.youtube.com/watch?v=EDauz38xV9w.

11. Stephen M. Stigler, *The History of Statistics: The Measurement of Uncertainty Before 1900* (Cambridge, MA: Harvard University Press, 1986), 64–65.

12. Since its publication in 1713, the Bernoulli theorem has gone through a series of upgrades.

13. For a proof, see Warren Weaver, *Lady Luck: The Theory of Probability* (Garden City, NY: Doubleday, 1963), 232–233.

14. Jacob Bernoulli, *The Art of Conjecturing*, trans. Edith Dudley Sylla (Baltimore: Johns Hopkins, 2006), 339.

15. Stigler, *The History of Statistics*, 77.

16. Bernoulli, *The Art of Conjecturing*, 329.

17. John Albert Wheeler, "Biographical Memoirs," vol. 51 (Washington, DC: National Academies Press, 1980), 110. The quote is a paraphrase of the original "God does not play dice," which appears in letters from Einstein to Max Born; see A. Einstein, *Albert Einstein und Max Born, Briefwechsel, 1916–1955, Kommentiert von Max Born* (Munich: Mymphenburg, 1969), 129–130.

18. Robert Oerter, *The Theory of Almost Everything* (New York: Pi Press, 2006), 84.

19. Mazur, *What's Luck Got to Do with It?*, 129–130.

20. Bernoulli, *The Art of Conjecturing*, 101.

21. There was one other major treatise on probability theory. In 1708 French mathematician Pierre Rémond de Montmort published his *Essai d'analyse sur les jeux de hazard* (Analytical Essay on Games of Chance).

22. Cardano's *Liber de Ludo Aleae* (Book on Games of Chance) was written in the 1500s and published in 1663, whereas Huygens's *De Ratiociniis in Ludo Aleae* (On Reasoning in Games of Chance) was published in 1657. However, the medieval poem "De Vetula," ascribed to Richard de Fournival, contained a short description of which combinations can come from the tossing of three dice without reference to any hint of expected value.

23. This quote appears on page 132 of Edith Dudley Sylla's translation of Bernoulli's *Ars Conjectandi*. Huygens's *De Ratiociniis in Ludo Aleae* is reproduced as Part 1 of *Ars Conjectandi*. It actually appeared first as an appendix to a book of mathematical exercises by Frans van Schooten, printed in 1657. Huygens's book is not to be confused with Girolamo Cardano's *mathematics* gambling manual, *Liber de Ludo Aleae*.

Chapter 6

1. Overall 3 percent of data were missing.

2. Victor Grech, Charles Savona-Ventura, and P. Vassallo-Agius, "Unexplained Differences in Sex Ratios at Birth in Europe and North America," *British Medical Journal* 324. no. 7344 (April 27, 2002).

3. Persi Diaconis, Susan Holmes, and Richard Montgomery, "Dynamical Bias in the Coin Toss," *SIAM Review* 49, no. 2 (2000): 211–235.

Chapter 7

1. Robert Siegel and Andrea Hsu, "What the Odds Fail to Capture When a Health Crisis Hits," NPR *All Things Considered*, July 21, 2014.

2. Road miles are according to the US Department of Transportation and Federal Highway Administration; square miles of land are according to the US Department of Agriculture Forest Service.

3. It may seem peculiar that in a hundred rounds of playing red in roulette, you are only likely to win forty-seven times and not fifty, but that comes from the fact that $p < q$ and so the peak probability is skewed away from the mean.

4. Mazur, *What's Luck Got to Do with It?*, 104.

5. However, to fit it on a page, the graph must be horizontally scaled down to look like the graph in Figure 7.4.

6. I am told that there are earlier reports of the triangle, starting with the work of the twelfth-century Indian mathematician Halaydha, who wrote a commentary on the Chandas Shastra (the Sanskrit treatise on the study of poetic meter), where he noticed that the diagonals of the triangle sum to what was later to be called the Fibonacci numbers. I have not seen corroborating evidence that such a triangle exists at such an early date, though

it may be so. If it does, it surely does not consider the formula for construction and may simply list enough rows to be useful.

7. Petrus Apianus was a German humanist, mathematician, and astronomer. See D. E. Smith, *History of Mathematics* (New York: Dover, 1958), 508.

8. Mazur, *What's Luck Got to Do with It?*, 239.

9. First, we shift the entire graph so that the high point is centered at 0. Area is clearly preserved and no information is lost, except that now we must interpret the meaning of the graph as the distribution of probabilities of the incremental increase or decrease of reds over blacks. For one further modification of our figure, we shrink the curve by a factor of 5 in the horizontal direction and magnify the curve by that same factor in the horizontal direction. The factor of 5 comes from the computation of $\sqrt{Npq}$, where N is the number of rounds, p is the probability of getting red, and q is the probability of not getting red. The precise number is 4.99307. I rounded it to 5 for the convenience of instruction.

10. First we had to rigidly shift the curve so its mean fell above 50. Then we had to compute a scalar (a scaling factor) by which to shrink the curve horizontally and magnify it vertically. The shift was a matter of knowing there were one hundred rounds of the game.

11. The scalar is $\sqrt{Npq}$, where N is the number of rounds, p is the probability of success, and q is the probability of failure ($q = 1 - p$). In other words, the scalar for our particular game of playing red in roulette is $\sqrt{100\left(\frac{9}{10}\right)\left(\frac{10}{9}\right)} = 4.99307$, or approximately 5.

12. The general picture of all the scaling and manipulation that we did can be seen as simple transformations of the variables x and y to new variables X and Y. We let $X = x–a$, to rigidly slide the original graph a units to the right. We let $X = x/b$, to

horizontally scale the original graph by a factor of b. Then finally, we let $Y = cy$, to vertically scale the original graph by a factor of c. In the end we have a new graph, Y vs. X. For a binomial frequency distribution with p relatively close to q, we may transform x into X by letting $= \frac{x-\left(\frac{N}{2}+Np+\frac{1}{2}\right)}{\sqrt{Npq}}$.

13. The curve depicted by the graph of $Y = \frac{1}{\sqrt{2\pi}}e^{-\frac{X^2}{2}}$ is called the *standard normal distribution*, which is really goes all the way back to de Moivre and Laplace. It is what you get from the normal distribution $y = \frac{1}{\sigma\sqrt{2\pi}}e^{-\frac{1}{2}\left(\frac{x-\mu}{\sigma}\right)^2}$ when $\mu = 0$ and $\sigma^2 = 1$ (μ is the mean and σ is the standard deviation).

14. Karl Pearson, *The Chances of Death and Other Studies in Evolution* (London: Edward Arnold, 1897), 45.

15. We are talking about roulette in Monaco. American roulette differs from European by including a double-zero slot as well as zero. However, the coin flipping analogy is very similar—the double zero counts as both red and black.

16. Pearson, *The Chances of Death and Other Studies in Evolution*, 55.

17. Ibid., 61.

18. Ibid., 55.

19. Warren Weaver, *Lady Luck, The Theory of Probability* (Garden City, NY: Doubleday, 1963), 282.

20. John Scarne, *Scarne's Complete Guide to Gambling* (New York: Simon & Schuster, 1961), 24.

Chapter 8

1. E. H. McKinney, "Generalized Birthday Problem," *American Mathematical Monthly* 73, (1966): 385–387.

2. Persi Diaconis gave an approximate fit; Bruce Levin's data points to this curve by the function $N \approx 47(k-1.5)^{3/2}$.

3. Richard von Mises, "Ueber Aufteilungs- und Besetzungs-Wahrscheinlichkeiten," *Review of Faculty of Science. University of Istanbul* 4 (1939), 145–163.

4. What is the probability $p(N)$ that one number will be picked twice after N picks? The answer is $p(N) = \prod_{k=1}^{N-1}\left(1 - \frac{k}{365}\right)$. To compute this we take the natural log of both sides to get $\ln(p(N)) = \sum_{k=1}^{N-1} \ln\left(1 - \frac{k}{365}\right)$. Since $\ln(1+x) \approx x$, we may approximate each kth term on the right side by $-k/365$, so the right side becomes approximately $\frac{1}{365}\sum_{k=1}^{N-1} k \approx -\frac{1}{365}\left(\frac{N(N-1)}{2}\right)$ which in turn is approximately $\frac{-N^2}{730}$, for large N. So, we know that $\ln(p(N)) \approx -\frac{N^2}{730}$. And if we solve for N, we get $N \approx \sqrt{2(365)(-\ln p(N))}$. In the case where $p = 1/2$, we get $N \approx 22.49$.

5. $N \approx \sqrt{2(9999)(-\ln(1/2))} \approx \sqrt{2(9999)(-\ln(1/2))} \approx 1.18\sqrt{9999} \approx 118$.

6. Take the natural logarithm of both sides of $\left(\frac{364}{365}\right)^N = \frac{1}{2}$ to get $N = \ln(1/2)/\ln(364/365) = 252.65$.

7. The equation to solve is $\left(\frac{7299}{7300}\right)^N = 1/2$. By taking the natural logarithm of both sides, we find that $N = \frac{\ln\left(\frac{1}{2}\right)}{\ln\left(\frac{7299}{7300}\right)} = 5{,}104.65$.

8. Sir Arthur Eddington, *The Nature of the Physical World*, (New York: Macmillan Company, 1927), 72.

9. The keys hits are independent; however, some hits might be more likely than others, given their position on the keyboard.

10. The graph of $P = (1 - (1/26)^5)^N$.

11. Émile Borel, "Mécanique Statistique et Irréversibilité," *Journal of Physics* series 5e, vol. 3 (1913): 189–196.

12. Sir James Jeans. *The Mysterious Universe* (New York: Macmillan, 1930), 4.

13. Darren Wershler-Henry, *The Iron Whim: A Fragmented History of Typewriting* (Ithaca, NY: Cornell University Press, 2007), 192.

Chapter 9

1. The typical board-game die has its dots gouged from the sides of a cube. Each gouge is as deep as the next, so the side with six gouges is lighter than the side with one gouge. Such a die is dishonest, as it favors heavier sides. To make an honest die, material gouged from one side should weigh the same as the material gouged from any other side. The paint to make the dots should also be weighed and balanced.

2. The uniformity will happen in the horizontal direction. Pressure differential creates a continuous gradation in the vertical direction, so it takes a longer time to see the vertical uniformity. Experiment with a relatively shallow bottle to create a better uniformity.

3. See Mark Kac, "Probability," *Scientific American*, September 1964.

4. Jacob Bernoulli, *The Art of Conjecturing*, trans. Edith Dudley Sylla (Baltimore: Johns Hopkins, 2006), 339.

5. William Paul Vogt and Robert Burke Johnson, *Dictionary of Statistics & Methodology: A Nontechnical Guide for the Social Sciences*, 4th ed. (Thousand Oaks, CA: SAGE Publications, 2011), 374.

6. Vogt and Johnson, *Dictionary of Statistics & Methodology*, 217.

7. Darrell Huff, *How to Lie with Statistics* (New York: Norton, 1993), 100–101.

8. Gary Taubes, "Do We Really Know What Makes Us Healthy?" *New York Times*, September 16, 2007.

9. J. H. Bennett, ed., *Statistical Inference and Analysis: Selected Correspondence of R. A. Fisher*, (Oxford: Oxford University Press, 1989).

10. Paul D. Stolley, "When Genius Errs: R. A. Fisher and the Lung Cancer Controversy," *American Journal of Epidemiology* 133, no. 5 (1991).

11. R. A. Fisher, *Collected Papers*, vol. 1, ed. J. H. Bennett (Adelaide, Australia: Coudrey Offset Press, 1974), 557–561.

12. Ronald A. Fisher (letters to *Nature*), "Cancer and Smoking," *Nature* 182, August 30, 1958.

13. Stolley, "When Genius Errs."

14. Sir Ronald Fisher, "Cigarettes, Cancer, and Statistics," *Centennial Review* 2 (1958): 151–166.

15. Marcia Angell and Jerome Kassirer, "Clinical Research—What Should the Public Believe?" *New England Journal of Medicine* 331 (1994), 189–190.

16. Taubes, "Do We Really Know What Makes Us Healthy?"

17. Samuel Arbesman, *The Half-Life of Facts: Why Everything We Know Has an Expiration Date* (New York: Current, 2012), 7.

Chapter 10

1. Woollcott, *While Rome Burns* (New York: Viking Press, 1934), 23.

2. Francesco is the third-most-frequent name in Italy, after Marco and Andrea. Manuela is not on the list of the first hundred most frequent names in Spain.

3. Actually, sixteen is a conservative multiplier, given that Maria, Laura, Marta, and Paula are far more frequently used names than Manuela.

4. It is not clear, from Flammarion's writing of this story, if the proof sheets were for the book he was working on or for some book that was already completed.

5. Joseph Mazur, *What's Luck Got to Do with It?: The History, Mathematics, and Psychology of the Gambler's Illusion* (Princeton, NJ: Princeton University Press, 2010), 177–178.

6. This is according to Nathanial Rich. See Nathanial Rich, "The Luckiest Woman on Earth," *Harper's Magazine*, August 2011. Rich's figure is off by almost a million times his calculation. The

correct odds are more than 2 nonillion to 1. (A nonillion is 1 followed by 30 zeros.)

Chapter 11

1. Warren Goldstein, *Defending the Human Spirit: Jewish Law's Vision for a Moral Society* (Jerusalem, Israel: Feldheim, 2006), 269.
2. J. Boyer, "DNA on Trial," *New Yorker*, January 17, 2000.
3. Michael R. Bromwich, head of investigating team, HPD Crime Lab Independent Investigation Report, May 11, 2006. Available at http://www.hpdlabinvestigation.org, accessed August 22, 2014.
4. Tobias Jones, "The Murder That Has Obsessed Italy," *The Guardian*, January 8, 2015.
5. William C. Thompson, Franco Taroni, and Colin G. G. Aitken, "How the Probability of a False Positive Affects the Value of DNA evidence," *Journal of Forensic Science* 48, no 1 (January 2003, 47–54.
6. Ibid., 47.
7. National Academy of Sciences (NAS) report, "Strengthening Forensic Science in the United States: A Path Forward" (2009).
8. Spencer S. Hsu, "D.C. Judge Exonerates Santae Tribble in 1978 Murder, Cites Hair Evidence DNA Test Rejected," *Washington Post*, December 14, 2012.
9. NAS, "Strengthening Forensic Science," 160.
10. Norman L. Reimer, https://www.nacdl.org/champion.aspx?id=29488.
11. See the Innocence Project piece on Santae Tribble at http://www.innocenceproject.org/cases-false-imprisonment/santae-tribble.
12. Brandon L. Garrett, *Convicting the Innocent: Where Criminal Prosecutions Go Wrong* (Cambridge, MA: Harvard University Press, 2011), 101.
13. NAS Report, 86.
14. Garrett, *Convicting the Innocent*, 101.

15. The set from the mother and the set from the father contain different versions of the same genes, the size of the genome is typically given as the number of bases in one set of genes.

16. The quote is from someone who had nothing to do with this case, Anita Alvarez, state's attorney for Cook County, Illinois.

17. Trisha Meili, *I Am the Central Park Jogger: A Story of Hope and Possibility* (New York: Scribner, 2004), 108.

18. Ibid., 6–7.

19. Jed S. Rakoff, "Why Innocent People Plead Guilty," *New York Review of Books* 61, no. 18, November 20, 2014, 16–18.

20. National Research Council Report, "The Growth of Incarceration in the United States" (2014).

21. Heather West, William Sabol, and Sarah Greenman, "Prisoners in 2009," US Department of Justice, Bureau of Justice Statistics, 2009, rev. October 27, 2011; Lauren E. Glaze and Erinn J. Herberman, "Correctional Populations in the United States, 2012," US Department of Justice, Bureau of Justice Statistics (2013), 2 and table 1, available at http://www.bjs.gov/content/pub/pdf/cpus12.pdf; Todd D. Minton, "Jail Inmates at Midyear 2012—Statistical Tables," US Department of Justice, *Bureau of Justice Statistics* 1 (2013), available in PDF format at http://www.bjs.gov/content/pub/pdf/jim12st.pdf.

22. The total federal and state criminal justice system spent $260,533,129,000 in 2010. This includes judicial and legal costs ($56.1 billion), police protection costs ($124.2 billion), and corrections costs ($80.24 billion).

23. Oliver Roeder, Lauren-Brooke Eisen, and Julia Bowling, "What Caused the Crime Decline?" Brennan Center for Justice at NYU School of Law, research report, 2015.

24. NAACP Legal Defense and Educational Fund, quarterly report by the Criminal Justice Project, Total number of death row inmates in US prisons as of January 1, 2014: 3,070; race of defendant: white 1,323, black 1,284, Latino/Latina 388, Native American 30, Asian 44.

25. NAACP Legal Defense Fund, "Death Row USA," January 1, 2014.

26. R. J. Maiman and R. J. Steamer, *American Constitutional Law: Introduction and Case Studies* (St. Louis, MO: McGraw-Hill, 1992), 35.

27. Cass R. Sunstein, "The Reforming Father," *New York Review of Books*, vol. 51, no. 10, June 5, 2014, 8.

28. Sources: US Department of Justice, Bureau of Justice Statistics, "Capital Punishment" for the years 1968–2012; NAACP Legal Defense and Educational Fund, Inc. "Death Row USA" for the years 2013 and 2014.

29. Sunstein, "The Reforming Father," 10.

30. Innocence Project report, "Reevaluating Lineups: Why Witnesses Make Mistakes and How to Reduce the Chance of a Misidentification" (2009), 17.

31. Garrett, *Convicting the Innocent*, 5.

32. Innocence Project, "Reevaluating Lineups," 5.

33. The National Registry of Exonerations of the University of Michigan Law School and the Center on Wrongful Convictions at Northwestern University School of Law; see http://www.law.umich.edu/special/exoneration/Pages/browse.aspx.

34. This was allegedly the case with Charles Hynes, the Brooklyn district attorney who was accused of such practices during the exoneration hearing of Jabbar Collins, the man who spent sixteen years in prison for a murder he did not commit. The wrongful damages were settled with New York City for $10 million. See Stephanie Clifford, "Exonerated Man Reaches $10 Million Deal with New York City," *New York Times*, August 19, 2014.

35. Goldstein, *Defending the Human Spirit*, 269.

Chapter 12

1. Pasteur Vallery-Radot, ed., *Oeuvres de Pasteur*, vol. 7 (Paris, France: Masson and Co., 1939), 131.

2. Gerard Nierenberg, *The Art of Creative Thinking* (New York: Simon & Schuster, 1986), 201.

3. Bruce W. Lincoln, *Sunlight at Midnight: St. Petersburg and the Rise of Modern Russia* (Boulder, CO: Basic Books, 2002), 150–151.

4. Victor E. Pullin and W. J. Wiltshire, *X-rays: Past and Present* (London: E. Benn Ltd., 1927).

5. Röntgen thought that X-rays are invisible. In, fact they can produce a blue-gray glow. See K. D. Steidley, "The Radiation Phosphene," *Vision Research* 30 (1990): 1139–1143.

6. W. R. Nitske, *The Life of Wilhelm Conrad Röntgen, Discoverer of the X Ray* (Tucson: University of Arizona Press, 1971).

7. Barbara Goldsmith, *Obsessive Genius: The Inner World of Marie Curie* (New York: W. W. Norton, 2005), 64.

8. Lawrence K. Russel, Poem, *Life*, 27, March 12, 1896.

9. Goldsmith, *Obsessive Genius*, 65.

10. Howard H. Seliger, "Wilhelm Conrad Röntgen and the Glimmer of Light," *Physics Today*, November 1995, 25–31.

11. "Fifty Years of X-Rays," *Nature*, 156, November 3, 1945, 531.

12. H. J. W. Dam, "The New Marvel in Photography," *McClure's Magazine* 6, no 5, April, 1896. *McClure's* folded for good in the 1929 crash. Fortunately, the Gutenberg Project has archived almost all of *McClure's* electronically.

13. J. McKenzie Davidson, "The New Photography," *The Lancet* 74, I (March 21, 1896): 795, 875.

14. *Nature* 53 (January 23, 1896): 274.

15. Otto Glasser, *Wilhelm Conrad Röntgen and the Early History of the Röntgen Rays* (San Francisco: Norman Publishing, 1993), 47–51.

16. *Atomic Physics*, film produced by the J. Arthur Rank Organization, 1948.

17. From the inaugural lecture of Louis Pasteur as professor and dean of the faculty of science, University of Lille, Douai, France,

December 7, 1854. See Houston Peterson, ed., *A Treasury of the World's Great Speeches* (New York: Simon and Schuster, 1954), 473.

18. Isaac Newton, *The Correspondence of Isaac Newton, Vol. 1. 1661–1675*, ed., H. W. Turnbull (Cambridge, UK: Cambridge University Press, 1959), 416.

19. John of Salisbury, *The Metalogicon: A Twelfth Century Defense of the Verbal and Logical Arts of the Trivium*, trans. Daniel McGarry (Baltimore: Paul Dry Books, 2009), 167.

20. Steven Weinberg, *Lake Views: This World and the Universe* (Cambridge, MA: Belknap Press, 2009), 187.

Chapter 13

1. B. F. Skinner's reason for the increased likelihood that the gambler will continue to play.

2. James B. Stewart, "The Omen," *New Yorker*, October 20, 2008, 58.

3. Ibid., 63.

4. Nelson D. Schwartz, "A Spiral of Losses by a 'Plain Vanilla' Trader," *New York Times* (January 25, 2008).

5. Nick Leeson, *Rogue Trader* (New York: Time Warner, 1997).

6. Russell Baker, "A Fateful Election," *New York Review of Books*, November 6, 2008, 4.

7. Seth Stein and Michael Wysession, *An Introduction to Seismology, Earthquakes, and Earth Structure* (Hoboken, NJ: Wiley-Blackwell, 2002), 5–6.

8. Florin Diacu, *Megadisasters: The Science of Predicting the Next Catastrophe* (Princeton, NJ: Princeton University Press, 2010), 29.

9. Charles Richter, "Acceptance of the Medal of the Seismological Society of America," *Bulletin of the Seismological Society of America* 67 (1977): 1.

Chapter 14

1. Michael Shermer, *Why People Believe Weird Things* (New York: Henry Holt, 1997), 69.

2. Elizabeth Gilbert, *The Signature of All Things* (New York: Viking, 2013), 483.

3. Actually, it was a Chinese laborer who discovered the frog and brought it to Wallace.

4. Luis A. Cordón, *Popular Psychology: An Encyclopedia* (Westport, CT: Greenwood, 2005), 182.

5. D. J. Bern and C. Honorton, "Does Psi Exist? Replicable Evidence for an Anomalous Process of Information Transfer," *Psychological Bulletin* 115 (1994): 4–8.

6. Lourdes Garcia-Navarro, "Letter from Beyond the Grave: A Tale of Love, Murder and Brazilian Law," National Public Radio News, *Weekend Edition*, August 9, 2014.

7. Martin Gardner, *Fads and Fallacies in the Name of Science* (New York: Dover, 1957), 299–307.

8. Stanton Arthur Coblentz, *Light Beyond: The Wonderworld of Parapsychology* (Vancouver: Cornwall, 1981): 109–110.

9. Sir Hubert Wilkens and Harold Sherman, *Thoughts Through Space: A Remarkable Adventure in the Realm of the Mind* (New York: Hampton Roads, 2004), 26–27.

10. Eric Lord, *Science, Mind and Paranormal Experience* (Raleigh, NC: Lulu, 2009), 210–211.

11. Gardner, *Fads and Fallacies*, 351.

12. J. B. Rhine and L. E. Rhine, "An Investigation of a 'Mind Reading' Horse," *Journal of Abnormal and Social Psychology* 23, no. 4 (1929): 449.

13. C. D. Broad, "The Relevance of Psychical Research to Philosophy," *Philosophy* 24, no. 91 (1949): 291–309.

14. Joseph Banks Rhine, *The New World of the Mind* (London: Faber and Faber, 1953), 80.

15. Originally published in Ronald Aylmer Fisher, *Design of Experiments* (London: Oliver and Boyd, 1937) but can more easily be found in Ronald Aylmer Fisher, *Statistical Methods, Experimental Design, and Scientific Inference* (Oxford: Oxford University Press, 1990), 11–18.

16. Fisher's essay is really meant to be about the design of experiments and the concern over subjective error, but here the story is used to point to the connection between mathematics and experiment.

17. Fisher, *Statistical Methods*, 12.

18. George R. Price, "Science and the Supernatural," *Science*, new series, 122, no. 3165 (August 26, 1955): 359–367.

19. H. Houdini, *A Magician Among the Spirits* (New York: Harper, 1924), 138.

20. Ecclesiastes 1:5–7

21. John Milton, *The Portable Milton*, ed. Douglas Bush (New York: Viking, 1961), 416–417.

22. Roald Dahl, *Charlie and the Chocolate Factory* (New York: Bantam, 1973), 137.

Chapter 15

1. Vladimir Nabokov, *Laughter in the Dark* (New York: New Directions, 2006).

2. Eugene Ionesco, *The Bald Soprano and Other Plays* (New York: Grove Press, 1958), 18.

3. Hilary P. Dannenberg, *Coincidence and Counterfactuality: Plotting Time and Space in Narrative Fiction* (Lincoln, NE: University of Nebraska Press, 2008), 90.

4. My poor translation of a line at the end of the second stanza from *Sir Gawain and the Green Knight*, trans. Brian Stone (New York: Penguin, 1974), 22.

5. *Sir Gawain and the Green Knight: A Middle-English Arthurian Romance* trans. Jessie Weston, trans. (London: David Nutt, 1898), available at http://d.lib.rochester.edu/camelot/text/weston-sir-gawain-and-the-green-knight.

6. Ibid.

7. Ibid.

8. *Sir Gawain and the Green Knight*, ed. William Raymond Johnson (Manchester, UK: Manchester University Press, 2004), 25.

9. Richard Boyle, "The Three Princes of Serendip," *Sunday [London] Times*, July 30 and August 6, 2000.

10. Dov Noy, Dan Ben-Amos, Ellen Frankel, *Folktales of the Jews, Vol. 1, Tales from the Sephardic Dispersion* (Philadelphia, PA: The Jewish Publication Society, 2006), 318–319.

11. The letter was to Horace Mann, not the American education reformist, but the British baronet and envoy to the court at Florence.

12. Robert K. Merton and Elinor Barber, *The Travels and Adventures of Serendipity: A Study in Sociological Semantics and the Sociology of Science* (Princeton, NJ: Princeton University Press, 2003), 3–4.

13. Boyle, "The Three Princes of Serendip."

14. *The Travels and Adventures of Three Princes of Sarendip* (London: William Chetword, 1722).

15. Other accounts of the same story appear in Idries Shah, ed. *World Tales: The Extraordinary Coincidence of Stories Told in All Times, in All Places* (London: Octagon, 1991), 336–339, and in Mrs. Howard Kingscote and Pandit Natesa Sastri, *Tales of the Sun or Folklore of Southern India* (Whitefish, MT: Kessinger Publishing, 2010 [originally published by W. H. Allen, 1890]), 140.

16. John Pier, and José Angel Garcia Landa, eds., *Theorizing Narrativity* (Berlin: Walter de Gruyter, 2007), 181.

17. Paul Auster, *Moon Palace* (New York: Viking, 1989), 236–237.

Epilogue

1. David Hand, *The Improbability Principle: Why Coincidences, Miracles, and Rare Events Happen Every Day* (New York: Farrar Straus and Giroux, 2014), 76. *Fluke* and *The Improbability Principle* are two very different books that approach the subject of coincidences from distinct perspectives that complement each other.

2. In 1980 physicist Luis Alvarez and his son geologist Walter Alvarez identified high concentrations of iridium in rock layers that mark the end of the Cretaceous period. The theory (a very controversial one) from the 1980s until 2013 was that an immense asteroid crashed into the earth at high impact. In 2013 Mukul Sharma and Jason Moore in the department of Earth Sciences at Dartmouth presented a paper at the 44th Lunar and Planetary Conference on the theory that it was not an asteroid, but rather a comet.

Acknowledgments

FIRST TO THANK is my wife, Jennifer Mazur. She has given me her unconditional support from the beginning, when she worried that this book might dilute the mystiques and charms of great stories. She is my strength, my resolve, and my early draft editor, a person who always gives me honest, brutal criticism followed by constructive advice for making things better.

The idea for writing this book was not mine. It came from dinner conversations at a residency for Bogliasco Foundation Fellows. For some unanticipated reason, conversations were repeatedly drawn to coincidence stories that swung between personal accounts, folklore yarns, fictional tales, and chronicles of accidental scientific discoveries. Each night I would think about whether I could mathematically explain the surprising frequency of coincidences. Each morning I would come to breakfast feeling that I was ready to explain them. By evening, my theories were in tatters, ready to be abandoned and replaced by more thoughtful arguments. Even so, my Bogliasco colleagues kept encouraging me to write a book about coincidences. So, I owe the inspiration of this book, first, to the Bogliasco Foundation, and second, to the happenstance of conversations with my delightfully inspiring resident companions, Anne-Marie Baron, David Heymann, Sandra Heymann, Paul Kane, Tina Kane, Liliana Menendez, Alistair Minnis, Florence Minnis, Helen Simoneau, Lewis Spratlin, and Melinda

Spratlin. They contributed more than they would admit to my enthusiasm for the topic.

A very special thanks to my caring manuscript readers: Jeffrey Bower, Michelle Bower, Deborah Clayton, Lewis Cohen, Sorina Eftim, Julian Ferholt, Deborah Ferholt, Nancy Heinemann, Tom Jefferies, Peter Meredith, Sam Northshield, Todd Smith, George Szpiro, and Jim Tober. Each contributed both directly and indirectly to the final draft of this book.

George Feifer, the author of *The Girl from Petrovka*, gave me the closest to any firsthand version of the famous Anthony Hopkins coincidence that I could get; I wrote to Anthony Hopkins directly and to his agent several times without getting a response. Francesco Marras, head of the Italian language school, Studitalia, gave me a full firsthand account of the Francesco/Manuela mistaken identity coincidence. Agnes Krup gave me the challenging problem of computing the likelihood that two people will bump into each other and learn that they share the same birth and year. Lisa Paolozzi told me about her twice-encounter with her albino taxi driver.

Very special appreciations go to my editors, TJ Kelleher and Ben Platt. Their meticulous readings, positive criticism, and intelligent editing suggested a restructuring that meaningfully clarified the book's central argument. To Quynh Do, associate editor at Basic Books, for her quick and intelligent responses to all my questions, and to my agent, Andrew Stuart, who saw this project's potential in my very brief proposal.

Index

JOSEPH MAZUR is emeritus professor of mathematics at Marlboro College and the author of several popular mathematics books including *Euclid in the Rainforest: Discovering Universal Trush in Logic and Math,* which was a finalist for the PEN/American Martha Albrand Award, and *The Motion Paradox: The 2,500-Year-Old Puzzle Behind All the Mysteries of Time and Space*. His most recent book is the highly acclaimed *Enlightening Symbols: A Short History of Mathematical Notation and Its Hidden Powers*. Among his many honors is a Guggenheim Fellowship, a Rockefeller Foundation Bellagio Residency, and a Bogliasco Fellowship. His writing has appeared in the *Wall Street Journal, New York Times, Guardian, Science, Nature,* and *Slate*. Mazur lives with his wife, Jennifer, in Marlboro, Vermont.